Antimicrobials in Environment

This book provides a multidisciplinary coverage of all manifestations of antimicrobials and antimicrobial resistance technology to promote eco-friendly processes and techniques for environmental sustainability. It covers various aspects of the multidisciplinary framework, applying principles of microbiology, environmental toxicology, and chemistry to assess the human and ecological risks associated with exposure to antibiotics or antibiotic resistance genes that are environmental contaminants. In addition, it also provides a variety of photographs, diagrams, and tables to help illustrate the material. Bringing together contributions from researchers on different continents with expertise in antibiotic resistance in a range of diverse environmental sections, the book offers a detailed reflection on the paths that make antibiotic resistance a global threat, and the state-of-the-art in antibiotic resistance surveillance and risk assessment in complex environmental conditions.

Students, researchers, scientists, environmentalists, academics, computational biologists, stakeholders, and policymakers can benefit from using *Antimicrobials in Environment* as a resource that addresses microbial biotechnology, microbiology, toxicology, and all disciplines related to antimicrobial research.

Features of the book:

- Covers antimicrobial resistance in the environment with up-to-date research.
- Includes recent references on each plausible antimicrobial resistance in the environment.
- Details the possible spread of antibiotic-resistant bacteria from an ecosystem.
- Describes the public health impact of the use of antibiotics in the environment.
- Presents cutting-edge research on nanotechnology, especially in food packaging, and emergent antimicrobial technologies.
- Highlights the antibiotic resistance in the environment: challenges and outlook.

Current Trends in Antimicrobial Research

Series Editor: Arti Gupta

Antimicrobials in Pharmaceutical and Medicinal Research
Edited by Arti Gupta and Ram Prasad

Antimicrobials in Food Science and Technology
Edited by Arti Gupta and Ram Prasad

Antimicrobial Resistance in Agriculture and its Consequences
Edited by Arti Gupta and Ram Prasad

Antimicrobials in Agriculture
Edited by Arti Gupta and Ram Prasad

Antimicrobials in Environment
Edited by Arti Gupta and Ram Prasad

For more info: Current Trends in Antimicrobial Research – Book Series – Routledge & CRC Press

Antimicrobials in Environment

Edited by
Arti Gupta and Ram Prasad

CRC Press
Taylor & Francis Group
Boca Raton London New York

CRC Press is an imprint of the
Taylor & Francis Group, an **informa** business

First edition published 2025
by CRC Press
2385 NW Executive Center Drive, Suite 320, Boca Raton FL 33431

and by CRC Press
4 Park Square, Milton Park, Abingdon, Oxon, OX14 4RN

CRC Press is an imprint of Taylor & Francis Group, LLC

Library of Congress Cataloging-in-Publication Data
Names: Gupta, Arti (Professor of biotechnology), editor. | Prasad, Ram
(Professor of microbiology), editor.
Title: Antimicrobials in environment / edited by Arti Gupta and Ram Prasad.
Other titles: Current trends in antimicrobial research.
Description: First edition. | Boca Raton FL : CRC Press, 2025. |
Series: Current trends in antimicrobial research | Includes bibliographical references and index.
Identifiers: LCCN 2024026745 (print) | LCCN 2024026746 (ebook) | ISBN 9781032215495 (hardback) | ISBN 9781032215525 (paperback) | ISBN 9781003268925 (ebook)
Subjects: MESH: Anti-Infective Agents--toxicity | Environmental Pollutants--toxicity | Environmental Exposure--adverse effects | Risk Assessment | Environmental Pollution--prevention & control | Sustainable Development
Classification: LCC RA566 (print) | LCC RA566 (ebook) | NLM QV 250 | DDC 362.1969/8--dc23/eng/20240813
LC record available at https://lccn.loc.gov/2024026745
LC ebook record available at https://lccn.loc.gov/2024026746

ISBN: 978-1-032-21549-5 (hbk)
ISBN: 978-1-032-21552-5 (pbk)
ISBN: 978-1-003-26892-5 (ebk)

DOI: 10.1201/9781003268925

Typeset in Times
by Deanta Global Publishing Services, Chennai, India

Contents

Acknowledgement

First and foremost, we would like to thank God for his never-ending grace, mercy, and provision during what ended up being one of the toughest times of our life.

We are indebted to the many people who helped bring this book to light. I wish to thank series editors Dr Vijai Kumar Gupta, Dr Marc Gutierrez, and Dr Bonita Glanville-Morris; Senior Production Editor Dr Jyotsna Jangra; Commissioning Editor Dr Renu Upadhyaya; and Taylor & Francis/CRC Press, for their generous assistance, constant support, and patience in initializing the series volume. The editors are very thankful to Vice Chancellor Faizabad University, Ayodhya, and Mahatma Gandhi Central University, Motihari, for their kind support and constant encouragement. Special thanks are due to well-wishers, family members, and friends.

Dr Arti Gupta owes a debt of gratitude to her wonderful husband, Mr Saket Agrawal, and their son, Reyansh. Their unrelenting support and motivation have been pivotal in inspiring her to undertake this endeavour, overcome all obstacles, and assemble all the pieces into a coherent whole. Dr Ram Prasad gives his special regards to his family for their constant invaluable support and, motivation to publish. We both also give special thanks to all faculty colleagues and senior faculties of our team.

Shri Avadh Raj Singh Smarak Degree College, Gonda, UP, India **Arti Gupta**
Mahatma Gandhi Central University, Motihari, Bihar, India **Ram Prasad**

Preface

Antibiotics, disinfectants, and bacteria resistant to these substances have been detected in environmental compartments such as wastewater, sediments, and soils. Antibiotics are released into the environment through their use in medicine and their employment as growth promoters in animal husbandry, fish farming, and other fields of concern. There is increasing concern about the growing resistance of pathogenic bacteria in the environment and their ecotoxic effects. Increasingly, antibiotic resistance is seen as an ecological problem. This problem includes both the environment of resistance genes and the resistant bacteria. Resistance genes are universally present in nature in pathogens, commensals, and environmental microorganisms. This contributed work demonstrates how the environment plays a pivotal role in the development of antimicrobial resistance traits in bacteria and the distribution of resistant microbial species, resistant genetic material, and antibiotic compounds.

This book explores the various facets of antimicrobes in the environment and their perceived impact on human health, including updated information about all the genera of animal farming and pathogenic microbes associated. The valuable information may be exploited by policymakers and various groups of experts, including academic and research personnel engaged in clinical and medical microbiology, along with all veterinary and medical practitioners. Researchers, scientists, healthcare and management practitioners will find this an essential resource on the legal use of antibiotics in animals and humans.

The first and second chapters by Preethy et al. review the environmental contamination and impact of urban effluents on the antimicrobial resistance of microbes, and antimicrobial resistance in aquaculture at the juncture of global warming, respectively. In Chapter 3, Annapoorani and Saranya describe sustainable antimicrobial finishes for textiles. Extensive studies on the integrated approach to the process design and experimental analysis of lab-scale moving bed biofilm reactor using eco-friendly carriers are discussed by Santosh Kumar in Chapter 4. In Chapter 5, Shedge and Sambrani highlight the antimicrobial agents in germicidal products: a threat to the environment and human health. In Chapter 6, Lodhi et al. emphasize the environment's role in antibiotic resistance development. In Chapter 7, Das et al. describe antibiotic resistance in the environment: challenges and outlook, and in Chapter 8, Priya and Thirumalaiyammal highlight antimicrobial resistance in the environment and its impact. In Chapter 9, Arora et al. focus on foodborne microbes in the antimicrobial properties of food nano-packaging. Finally, in Chapter 10, Steffi et al. discuss applications of nanoparticles as antimicrobial agents.

Arti Gupta and Ram Prasad

Editors

Arti Gupta obtained her PhD from Mahatma Jyotiba Phule Rohilkhand University, Bareilly, India, in 2010 in Animal Science. Dr Gupta's current research interests are animal biotechnology, molecular plant biotechnology, molecular animal biotechnology, bioprocess technology, and microbiology. From July 2014 to date, Dr Gupta has been serving at Shri Avadh Raj Singh Smarak Degree College, Gonda, Uttar Pradesh, India, and has been engaged in some editorial work with Springer Nature, CRC Press, and De Gruyter. Dr Gupta has 12 years of teaching and research experience in Plant and Animal Biotechnology and Microbial Biotechnology. Dr Gupta has worked as external examiner for central evaluation in 2023–2024 at MJP Rohilkhand Univeristy, Bareilly.

Dr Gupta has received the following awards: University Topper (Gold Medal), MSc (Biotechnology) from ChCS University, Meerut; Young Scientist Award (Gold Medal) from the Zoological Society of India, Lucknow; Best Poster Presenter from *Asian Journal of Experimental Science*, Jaipur; Best Poster Presenter from International Consortium of Contemporary Biologists (ICCB) and Madhawai-Shyam Educational Trust, Ranchi; Fellowship Award from International Consortium of Contemporary Biologist (FICCB) and Madhawi Shayam Educational Trust (FMSET); and Dr V.P. Agarwal (Gold Medal) from DAV (PG) College, Muzaffarnagar. She has been involved with the following publications: 2 monographs, 13 edited books with Springer Nature, Switzerland, CRC Press, T&Y USA, and GmBH Germany; 7 book chapters, and around 50 national and international research papers. She has presented 28 abstracts in national and international symposia/seminars/conference/workshops. Dr Gupta has lifetime membership of the Indian Science Congress Association, Biotech Research Society of India, Zoological Society of India, and International Consortium of Contemporary Biologists. She has been serving as a member on several editorial boards. Previously, Dr Gupta served as Teaching Personnel, GB Pantnagar University, Pantnagar, Uttarakhand, India; Visiting Academic Researcher, Biorefining and Advanced Material Research Center (BAMRC), Scotland's Rural College (SRUC), University of Edinburgh, Scotland, United Kingdom.

Ram Prasad, PhD, is associated with the Department of Botany, Mahatma Gandhi Central University, Motihari, Bihar, India. His research interests include applied and environmental microbiology, plant microbe – interactions, sustainable agriculture, and nanobiotechnology. Dr Prasad has more than 300 publications to his credit, including research papers, review articles, and book chapters, and 7 patents issued or pending, and he had edited or authored several books. He has guided one postdoc, two MPhils, nine PhDs, and several master dissertation students. Dr Prasad has 14 years of teaching/research experience and has been awarded the Sir CV Raman Scientist Award, Young Scientist Award & Prof. J.S. Datta Munshi Gold Medal by the International Society for Ecological Communications; Fellow of Agricultural Technology Development Society (FATDS); Fellow of Biotechnology Research Society of India (FBRS); Fellow of the Society for Applied Biotechnology (FSAB); the American Cancer Society UICC International Fellowship for Beginning Investigators, USA; Outstanding Scientist Award in the field of Microbiology; BRICPL Science Investigator Award, Research Excellence Award, etc. He has been serving as a member of several editorial boards. Previously, Dr Prasad served as Assistant Professor at Amity University Uttar Pradesh, India; Visiting Assistant Professor, Whiting School of Engineering, Department of Mechanical Engineering at Johns Hopkins University, Baltimore, United States, and Research Associate Professor at the School of Environmental Science and Engineering, Sun Yat-sen University, Guangzhou, China.

Contributors

Anand Barapatre
Central Instrumentation Facility
Faculty of Science
Indira Gandhi National Tribal University
Amarkantak, Madhya Pradesh, India

Anchal Lodhi
Kumaun University
Nainital, Uttarakhand, India
Somaiya Vidyavihar University
Mumbai. Maharashtra, India

B. Thirumalaiyammal
Department of Microbiology
Cauvery College for Women (A)
Trichy, Tamil Nadu, India
Department of Microbiology
Indra Ganesan College of Arts And Science
Thiruchirappalli, Tamil Nadu, India

Bikas Kumar
Department of Biosciences and Bioengineering
Indian Institute of Technology
Guwahati, Assam, India
Biosciences and Biomedical Engineering
Indian Institute of Technology
Indore, Madhya Pradesh, India

Deepak Kumar Arya
Kumaun University
Nainital, Uttarakhand, India
Somaiya Vidyavihar University
Mumbai. Maharashtra, India

G. Sivaranjani
Department of Biotechnology
St. Joseph's College of Engineering
Chennai, Tamil Nadu, India

Harit Jha
Department of Biotechnology
Guru Ghasidas Vishwavidyalaya
Bilaspur, Chhattisgarh, India

K. R. Preethy
Department of Biotechnology
St. Joseph's College of Engineering
Chennai, Tamil Nadu, India

K. Hemavarshini
Department of Biotechnology
St. Joseph's College of Engineering
Chennai, Tamil Nadu, India

K. R. Preethy
Department of Biotechnology
St. Joseph's College of Engineering
Chennai, Tamil Nadu, India

K. Aishwarya
Department of Biotechnology
St. Joseph's College of Engineering
Chennai, Tamil Nadu, India

K. Saranya
Department of Textiles and Apparel Design
Bharathiar University
Coimbatore, Tamil Nadu, India

M. Chamundeeswari
Department of Biotechnology
St. Joseph's College of Engineering
Chennai, Tamil Nadu, India

Nidhi V. Maheshwari
Kumaun University
Nainital, Uttarakhand, India
Somaiya Vidyavihar University
Mumbai, Maharashtra, India

Nishant Burnase
Molecular Virology Lab
Chhattisgarh Institute of Medical Sciences (CIMS)
Bilaspur, Chhattisgarh, India

Nisha Suryawanshi
Department of Zoology
Government Arts and Commerce College
Sagar, Madhya Pradesh, India

N. Sathammai Priya
Department of Microbiology
Indra Ganesan College of Arts And Science
Thiruchirappalli, Tamil Nadu, India

P. F. Steffi
Department of Microbiology
Indra Ganesan College of Arts and Science
Thiruchirappalli, Tamilnadu, India
Department of Microbiology
Cauvery college for women (A)
Trichy, Tamil Nadu, India

R. Santhosh Kumar
Department of Civil Engineering
Vinayaka Mission's Kirupananda Variyar Engineering College, Vinayaka Mission's Research Foundation (DU)
Salem, Tamil Nadu, India

Reena Das
Dr. APJ Abdul Kalam Centre of Excellence in Innovation and Entrepreneurship
Dr. MGR Educational and Research Institute
Chennai, Tamil Nadu, India

R. Thenmozhi
Department of Microbiology
Cauvery College for Women (Autonomous)
Thiruchirappalli, Tamilnadu, India

Ragini Arora
Department of Biotechnology
Guru Ghasidas Vishwavidyalaya
Bilaspur, Chhattisgarh, India

S. L. Divya Praba
Department of Biotechnology
St. Joseph's College of Engineering
Chennai, Tamil Nadu, India

S. Grace Annapoorani
Department of Textiles and Apparel Design
Bharathiar University
Coimbatore, Tamil Nadu, India

Seema Sambrani
Biotechnology Department
S. K. Somaiya College
Somaiya Vidyavihar University
Mumbai, Maharashtra, India

Thenmozhi Manivel
Department of Microbiology
Indra Ganesan College of Arts and Science
Trichy, Tamil Nadu, India

Vaishnavi Shedge
Department of Biotechnology
S. K. Somaiya College
Somaiya Vidyavihar University
Mumbai, Maharashtra, India

1 Environmental Contamination and Impact of Urban Effluents on the Antimicrobial Resistance of Microbes

K.R. Preethy, M. Chamundeeswari, K. Hemavarshini, and S. L. Divya Praba

1.1 INTRODUCTION

Nature minimizes hazards, while man maximizes them. Environmental contamination occurs when resources are used irrationally in an inappropriate location and time. Urbanization, industrialization, and the excessive use of agrochemicals are potential sources of this contamination. Urban effluents are mainly from domestic and industrial sources. This effluent is composed of both organic and inorganic compounds along with a huge quantity of microorganisms. Antibiotics are typically used to treat these microorganisms. However, due to the continuous use of certain antibiotics, they have developed resistance to these antibiotics and are thus known as antimicrobial resistant microorganisms (AMRs). Domestic and industrial sources include sewage and drains from houses, effluent from factories, excessive use of fertilizers and pesticides on agricultural land, etc.

1.2 SOURCES OF ENVIRONMENTAL CONTAMINATION

A wide variety of chemicals can contaminate our water, land, or air, negatively impacting both the environment and our health. The majority of contaminants enter the environment through industrial and commercial facilities, oil and chemical spills, non-point sources like roads, parking lots, storm drains, wastewater treatment plants, and sewage systems ("Contaminants in the Environment").

Natural or xenobiotic (man-made) contaminants can exist in nature. Polyaromatic hydrocarbons, heavy metals, pesticides, organic and inorganic solvents, and chlorofluorocarbons are examples of such contaminants ("Environmental Contaminants: Sources and Effects"). Heavy metals like cadmium cause kidney damage and bone disease, and it is also classified as carcinogens (Godt et al. 2006), mercury leads to the central nervous system and cardiovascular disease (Azevedo et al. 2012), nitrates might be responsible for colon cancer in some individuals (Van Grinsven et al. 2006), and other pollutants are common environmental contaminants. The unintentional release of these contaminants into the environment causes the emergence of new diseases that endanger human health and result in mass population deaths ("Environmental Contaminants: Sources and Effects").

DOI: 10.1201/9781003268925-1

1.3 IMPACTS OF URBAN EFFLUENTS

1.3.1 Environment

The most direct impact of wastewater on the environment is when it aids in the deterioration and contamination of natural ecosystems and the wildlife that inhabits them by exposing them to dangerous substances that would not otherwise be present in the course of natural events.

1.3.2 Agriculture

Wastewater is frequently treated and used for various purposes, including irrigation. When wastewater is not properly handled, chemicals that are hazardous to crops may end up in the soil. The soil will produce minimal yield at a slower rate as a result of these contaminants. A further factor to take into account is the eventual consumption of these crops, which may be harmful to people.

1.3.3 Living Organisms

The ultimate consumption of these crops could be detrimental to people. Wastewater may contain heavy metals, pathogens, salts, toxic chemicals, oil and grease, solids, nutrients, sludge, acids and bases, hazardous organic compounds, organic and inorganic components, and nutrients. This effluent presents several risks to people, animals, and the environment as a whole. It may be flammable, corrosive, reactive, hazardous, and/or acidic. As a consequence, it must be treated before being reused or placed into the water system. In general, waterways are especially susceptible to the negative consequences of waste. Aquatic habitats are disturbed by toxic substances in the wastewater. Organisms begin breaking down enormous amounts of biodegradable materials that have penetrated the water, which requires a lot of dissolved oxygen. For marine life to thrive, dissolved oxygen levels are essential, and when they drop, fish may face a significant threat to their existence. Additionally, grease and oil in wastewater are harder to degrade and have a tendency to float on the surface of the water. This obstructs the light that aquatic photosynthetic plants require. It has the potential to entangle birds' feathers and choke fish. These are toxic to both humans and animals, just like heavy metals such as lead and mercury. Serious health consequences could also result from consuming fish that were captured in a contaminated water source, either as food or drink. The same issue applies to aquatic flora and animals. Untreated wastewater is frequently warm or even hot when dumped, which can elevate the water's temperature and further affect the ecology. Fish rely on water to maintain their body temperature because they have cold blood. The warmth of the water can make animals move more rapidly and breathe more heavily. The amount of oxygen in the water is also impacted by its temperature (2024).

1.4 DISEASES CAUSED BY URBAN EFFLUENTS

The following are some of the diseases caused by the pathogenic microbial community in urban effluents:

Campylobacteriosis – It is a diarrhoeal illness caused by the bacterium *Campylobacter*. Consuming contaminated water is one of the recognized sources of infection, and surface waterways frequently carry *Campylobacter* sp. (Rechenburg and Kistemann 2009).

Symptoms: Some people exhibit no symptoms; clinical manifestations include bloody diarrhoea, cramping, abdominal pain, nausea, vomiting, and fever within two to five days after exposure to the organism.

Cryptosporidiosis – It is a disease caused by the microscopic parasite *Cryptosporidium parvum*. It is very resistant to chlorine disinfection.

Symptoms: Some people exhibit no symptoms; most experience diarrhoea, loose or watery stools, stomach cramps, upset stomach, and a slight fever.

***Escherichia coli* diarrhoea** – It is also known as "diarrhoeagenic *E. coli*", a disease caused by *Escherichia coli* bacteria. This syndrome occurs particularly in children under the age of 5, the elderly, and those with weakened immune systems.

Symptoms: Some people exhibit no symptoms; most experience watery or bloody diarrhoea, abdominal cramps, nausea, vomiting, and fever.

Encephalitis – It is an inflammatory viral disease that is transmitted via the bite of infected mosquitoes, primarily of the *Culex* species.

Symptoms: Include high fever with head and body aches, stiff neck, muscle weakness, disorientation, tremors, convulsions, and, in the most severe cases, coma or paralysis. The illness can be severe for infants, the elderly, and those who are immuno-compromised. Vaccines are not yet produced for encephalitis.

Gastroenteritis – Viral gastroenteritis, also known as "stomach flu", is an infection caused by any number of viruses including rotaviruses, adenoviruses, caliciviruses, astroviruses, Norwalk virus, and a group of Norwalk-like viruses.

Symptoms: Include watery diarrhoea, vomiting, headache, fever, and abdominal cramps.

Salmonellosis – It is a disease caused by a group of bacteria called *Salmonella*, the most common of which are *Salmonella Typhimurium* and *Salmonella enteritidis.*

Symptoms: Include diarrhoea, fever, and abdominal cramps 12 to 72 hours after infection.

Shigellosis – Shigellosis is caused by a group of bacteria called *Shigella.*

Symptoms: Some people experience no symptoms, and most develop bloody diarrhoea, fever, and stomach cramps starting a day or two after exposure.

Typhoid fever – It is a disease caused by the bacterium *Salmonella typhi*, which lives only in the bloodstream and intestinal tract of humans.

Symptoms: Include sustained fever as high as 104°F, weakness, cough, stomach pains, headache, and loss of appetite.

To prevent us from contracting these infectious diseases, different antibiotics were used, which are discussed in the following section ("Diseases Involving Sewage" 2022).

1.5 ANTIBIOTICS AND THEIR MECHANISM

An antibiotic is a chemical compound that is harmful to other microorganisms and is produced by a living organism. In a complex environment like soil, organisms undoubtedly utilize antibiotics to regulate the growth of competing microbes. Antibiotics are frequently generated by soil microorganisms. Bacteria and fungi are microorganisms that produce antibiotics that are beneficial in preventing or treating disease. The mechanism of various groups of antibiotics is given in Figure 1.1. Bactericidal and bacteriostatic antibiotics are the two main classifications based on their mechanism of action: Bactericidal antibiotics work by killing the bacteria, while bacteriostatic antibiotics work by inhibiting the growth of the bacteria. Based on the mechanism of antimicrobial activity, antimicrobial agents are classified into the following groups (Munita and Arias 2016). The main groups are as follows: agents that inhibit cell wall synthesis, agents that depolarize the cell membrane, agents that inhibit protein synthesis, agents that inhibit nucleic acid synthesis, and agents that inhibit metabolic pathways in bacteria. Some of the most commonly found microorganisms in urban effluents are *Escherichia* sp., *Shigella* sp., *Salmonella* sp., and *Yersinia* sp. The acquisition of foreign DNA coding for resistance mechanisms through horizontal gene transfer (HGT) includes conjugation, transduction, and transformation, and mutations in a gene(s) consistently related to the mechanism of action of the chemical. Fluoroquinolones, norfloxacin, beta-lactams, macrolides, sulfamethoxazole, and trimethoprim are the most commonly encountered antibiotics. Antimicrobial-resistant

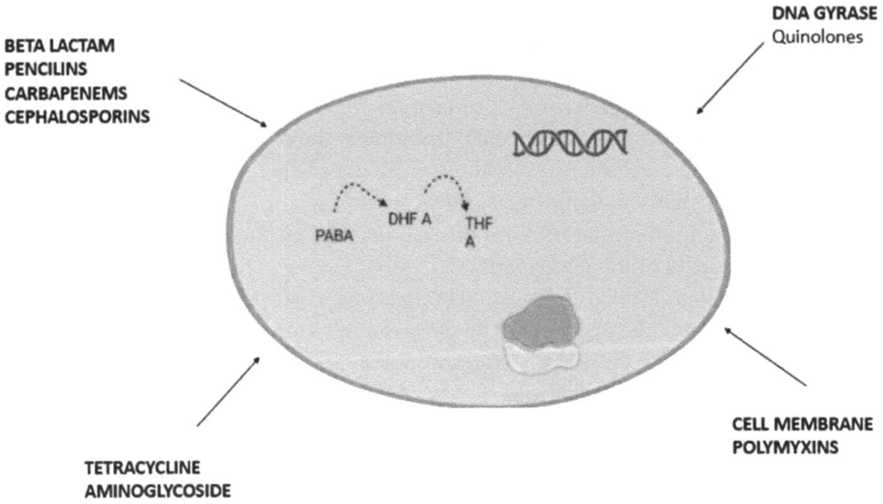

FIGURE 1.1 Mechanism of various groups of antibiotics on microorganisms.

microorganisms result from the overuse of these antibiotics on microorganisms (AMR) (Munita and Arias 2016).

1.5.1 Antimicrobial Resistance

One of the biggest threats to human health today is the emergence of resistance among the most significant bacterial pathogens. Multidrug-resistant organisms have not only become prevalent in hospitals but are also frequently observed in community settings, indicating that there are reservoirs of bacteria that are resistant to antibiotics beyond the medical facilities. The apex of evolution and the best-known example of bacterial adaptability is the bacterial response to the antibiotic attack. "Survival of the fittest" is a result of the enormous genetic plasticity of bacterial pathogens, which causes specific responses to lead to mutational adaptations, the acquisition of genetic material, or changes in gene expression, producing resistance to almost all antibiotics currently used in clinical practice (Davies, Julian, and Dorothy 2010) (Table 1.1).

1.5.2 Types of Antimicrobial Resistance

Natural Resistance

It can either be induced or intrinsic (always present in the species and the genes are naturally occurring in the bacteria, but are only expressed to resistance levels after exposure to an antibiotic). Intrinsic resistance refers to a trait that is present in all bacterial species regardless of previous antibiotic exposure or horizontal gene transfer. The reduced permeability of the outer membrane and the normal activation of efflux pumps are the two most frequent bacterial mechanisms implicated in intrinsic resistance. Another typical mechanism of induced resistance is the multidrug-efflux pump. It is possible for bacteria to acquire genetic material that imparts resistance by transformation, transposition, and conjugation (all referred to as horizontal gene transfer, or HGT), as well as through alterations to their own chromosomal DNA. The acquisition could be either temporary or permanent. The most prevalent way to acquire new genetic material is by the plasmid-mediated transmission of resistance genes; bacteriophage-borne transmission is rather uncommon. Some bacteria, including *Acinetobacter* species, are naturally competent and can thus take genetic material readily from their environment. Internally, insertion sequences and integrins may transfer genetic

TABLE 1.1
Antimicrobial Mechanism of Action of Various Antimicrobial Groups

Mechanism of Action	Antimicrobial Groups
Inhibit cell wall synthesis	Beta-lactams
	Carbapenems
	Cephalosporins
	Penicillin
Inhibit protein synthesis	Aminoglycosides
Binds to 30s ribosomal subunit	Tetracyclines
Binds to 50s ribosomal subunit	Chloramphenicol
	Macrolides
Inhibit deoxyribonucleic acid	Nalidixic acid
Inhibit nucleic acid synthesis	Quinolones
	Fluoroquinolones

information around, and stresses (starvation, UV radiation, toxins, etc.) on the bacteria are common causes of genetic alterations (substitutions, deletions, etc.) (Reygaert, Wanda C 2018).

Acquired Resistance

It is when a certain bacterium acquires the capacity to thwart an antimicrobial agent's effects to which it was previously vulnerable; this is known as acquired resistance. The acquisition of invading resistance genes, the mutation of genetic materials involved in regular physiological functions and cellular structures, or a combination of both methods can cause this. A successful gene transfer that could comprise mutation, horizontal gene transfer through transformation, transduction, or conjugation results in acquired resistance. In addition to plasmids, transposons, gene cassettes, integrative and conjugative elements, and other mobile elements, acquired resistance genes encoding for any of the three main resistance mechanisms – enzymatic inactivation, reduced intracellular accumulation, and modification of the cellular target sites – are also associated with these elements (Reygaert, Wanda C. 2018).

1.5.3 Microorganisms That Exhibit Resistance to the Most-Prominent Antibiotics

New strategies to tackle the rapidly spreading challenge of drug-resistant bacterial pathogens are desperately needed. These infections pose a serious public health threat.

1.5.3.1 Beta-Lactam

One of the most prominent classes of antibiotics, beta-lactams, is confronted by conflicts with resistance, and the emergence of new and more effective beta-lactam antibiotics by side chain modification of existing beta-lactam classes is not sustained by the emergence of resistance. Beta-lactam antibiotics include penicillins, cephalosporins, carbapenems, and monobactams (Worthington and Melander 2013). The structure of a beta-lactam is given in Figure 1.2.

1.5.3.2 Methods to Prevent Antibiotic Resistance in Beta-Lactam

Several bacterial strains have evolved resistance to each and every beta-lactam antibiotic (as well as every antibiotic in general), despite extensive pharmacological initiatives to alter beta-lactam drugs. One of two pathways is frequently liable for resistance to beta-lactam antibiotics. The two most prevalent forms of resistance in Gram-negative bacteria are:

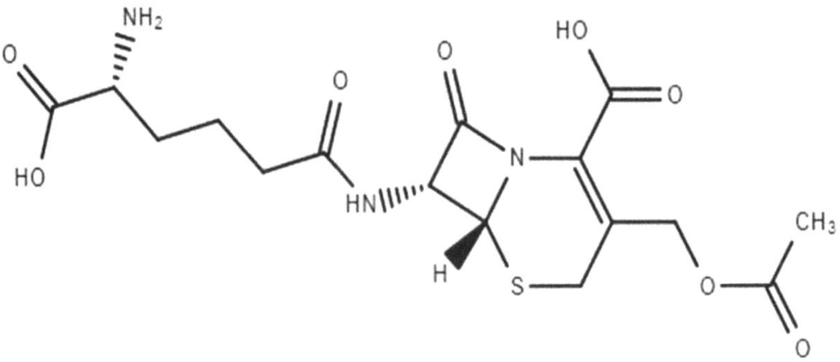

FIGURE 1.2 CaptionChemical structure of Beta-Lactam.

(1) Production of beta-lactamases; or
(2) Production of a modified PBP with a decreased affinity for most lactam antibiotics. Serine beta-lactamases and metallo-lactamases are the two structural subtypes of beta-lactamases (MBL).
1. There are essentially two possibilities for continuing to use beta-lactam antibiotics depending on these resistance mechanisms:
(1) Create brand-new beta-lactam antibiotics that are truly immune to the resistant bacteria mechanisms discussed above; or
2) Combine current beta-lactam antibiotics with a medication that neutralizes the resistance mechanisms.

Even while new beta-lactam antibiotics are still being researched, no distinct beta-lactam class has indeed been identified in more than 30 years, and new beta-lactam variants utilizing existing beta-lactam scaffolds only make up a small portion of the novel antibiotics now under development. Therefore, it is anticipated that in the future, combination therapy will be the most popular method of using beta-lactam antibiotics. The beta-lactam antibiotic class has undoubtedly been among the most significant and effective therapeutic classes for the past 70 years, and while it is not the only treatment option, it can still be quite helpful in the fight against illnesses brought on by pathogenic bacteria. The creation of new beta-lactam antibiotics would regrettably not be sufficient to keep up with the bacterial resistance that is continually changing. So, in our changing approach to treating multi-drug-resistant bacterial infections, adjuvant strategies to boost the efficacy of these antibiotics, such as those presented in this perspective, will probably be crucial (Bush and Bradford 2016) (Table 1.2).

1.5.3.3 Quinolones

Quinolones are a category of synthetic, broad-spectrum antibiotics that function by inhibiting DNA gyrase as well as topoisomerase IV. This causes DNA fragmentation and cell death as a consequence of genotoxic damage. The structure of levofloxacin (a class of quinolones) is given in Figure

TABLE 1.2
Examples of Bacteria with Intrinsic Resistance

Organisms	Intrinsic Resistance
Bacteroides (anaerobes)	Aminoglycosides, many β-lactams, quinolones
Enterococci	Aminoglycosides, cephalosporins, lincosamides

1.3. There are three major ways in which bacteria acquire quinolone resistance (SG Solano-Gálvez, MF Valencia-Segrove, 2020).

1.5.3.4 Mechanism of Quinolones

The enzyme DNA gyrase exists only in bacteria. This enzyme inserts negative supercoiling in DNA by using energy from ATP hydrolysis. Before strand transit, the DNA is chirally wrapped around this specialized domain of the enzyme, which results in the unidirectional supercoiling activity (Hooper and Jacoby 2015). Chromosome condensation, torsional strain relief during replication, and local melting for critical activities like RNA polymerase transcription initiation are all made possible by negative DNA supercoiling. Because DNA gyrase is required for bacterial development but is absent from eukaryotic cells, it is a great target for quinolones. Two subunits of this enzyme, A (97 kDa) and B (90 kDa), combine to produce an A2B2 tetramer. While the gyrB gene encodes the B subunit, which has ATPase activity, the gyrA gene encodes the A subunit, which is primarily engaged in DNA fragmentation and reunion. The separation of these two DNA ends creates a temporary gate, via which the subsequent DNA segment that has been coiled around DNA gyrase would then be transmitted. The distinctive negatively supercoiling ability of DNA gyrase in this process is due to the C-terminus of the GyrA subunit. The discovery that mutants devoid of the C-terminus are unable to produce negative supercoils has led to this conclusion. As previously suggested, quinolone drugs work by inhibiting synthesis, cell division, and DNA replication in order to suppress Type II topoisomerases (Morais-Cabral et al. 1997). The topoisomerase enzyme and DNA produce a ternary cleavage complex, which is the mechanism through which quinolones are inhibited. However, the chemical specifics of the mode of action of these medications are still unknown.

1. Chromosome changes in genes that code for proteins (mutations that alter the objectives of drugs).
2. Quinolone intracytoplasmic concentration-related mutations.
3. Quinolone resistance genes mediated by plasmids (plasmids that protect cells from quinolone toxicity).

1.5.3.5 Tetracycline

Tetracycline resistance is attributed primarily to one or more of the following: chromosomal mutations that enhance the expression of intrinsic resistance mechanisms, the acquisition of mobile genetic components carrying tetracycline-specific resistance genes, and/or mutations within the ribosome binding site (TH Grossman 2016). The antibiotic tetracycline prominently binds to the bacterial ribosome and interacts with a 16s ribosomal RNA (rRNA) target in the 30s ribosomal subunit, stopping translation by obstructing the docking of aminoacyl-tRNA during elongation.

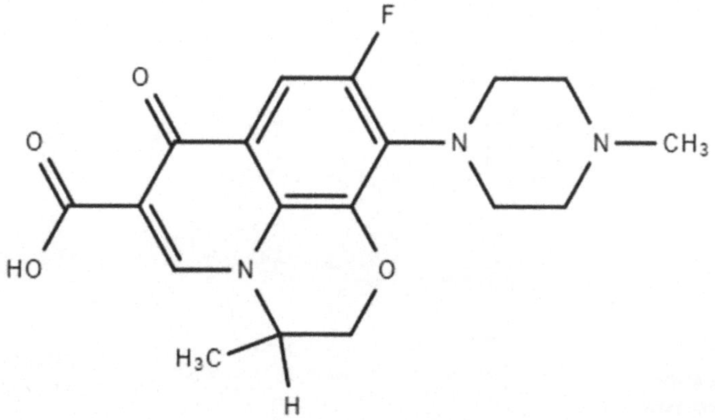

FIGURE 1.3 Structure of levofloxacin.

Many of the tetracycline ()(Roberts, Marilyn C.) ribosomal protection genes are found in both Gram-positive and Gram-negative bacteria, and they are widely distributed throughout the bacterial population on mobile genetic elements. These proteins catalyze tetracycline to be released from the ribosome in a GTP-dependent manner. Ribosomal protection proteins (RPPs) (Connell, Tracz, et al. 2003), which are considered to detach tetracycline from its binding site by interacting directly with the stacking interaction of the tetracycline D-ring and 16s rRNA, according to structural investigations using cryo-electron microscopy. RPPs and EF-G are thought to compete for an overlapping binding site. Resistance to tetracycline, minocycline, and doxycycline is conferred by RPP pathways. A simplified mechanism of tetracycline is given in Figure 1.4. Tet(O) is a protein that resembles an elongation factor that acquires resistance to tetracycline, an antibiotic that inhibits protein synthesis, by encouraging the drug's release from the inhibitory region on the ribosome (Connell, Trieber et al. 2003). In isolates bearing each of the three resistant strains of *Streptococcus* sp., the cause of tetracycline resistance was examined. Tetracycline uptake was evaluated in cells that were simultaneously tetracycline-sensitive and in cells that possessed any one of the three resistance genes. Uptake did not occur in cells that have tetL. However, tetracycline accumulated roughly 25-fold against a concentration gradient in sensitive cells and in cells that contained either tetM or tetN. Additionally, there was no proof that the intracellular tetracycline retrieved from sensitive tetM, or tetN cells had been altered. Tetracycline was able to inhibit protein synthesis in isolates made from organisms expressing tetM or tetN. Tetracycline, on the other hand, was sensitive to tetL cells (Burdett 1986).

1.6 METHODS TO INHIBIT ANTIBIOTIC RESISTANCE

There are numerous microorganisms developing resistance against antibiotics. Hence, there is a current pressing need to develop methods that would inhibit the antimicrobial resistance of microorganisms towards their corresponding antibiotics. There are various methods that can contribute to inhibiting antimicrobial resistance, such as

1. Essential oils
2. Gene modifications of integrons

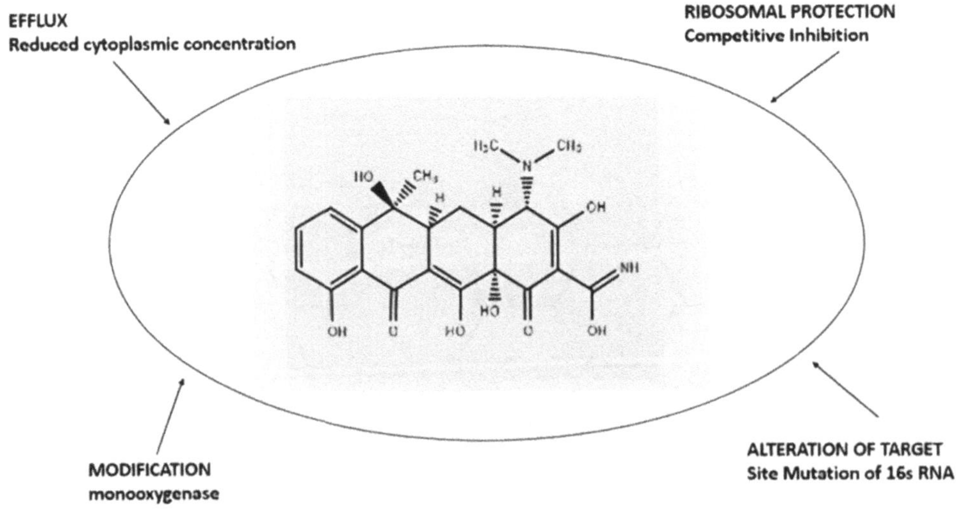

FIGURE 1.4 Mechanism of tetracycline.

1.6.1 Essential Oils

Essential oils are a complex mixture of volatile organic compounds. These oils possess antibacterial, antifungal, and antiviral activity. Additionally, they possess drug-resistant microbial strains. The conventional methods of antibiotics find it difficult to treat drug-resistant microbial strains (Tariq et al. 2019).

1.6.2 Mode of Action of Essential Oils in Various Microorganisms

1.6.2.1 Fungi

Essential oils establish a membrane potential across the cell wall and disrupt ATP assembly, leading to cell wall damage. Essential oils can also disintegrate the mitochondrial membrane, interfering with the electron transport system (ETS) pathway (Tariq et al. 2019).

1.6.2.2 Bacteria

Essential oils adversely impact several cellular processes in bacterial infections, including membrane transport and energy production, by destabilizing the cellular architecture. Essential oils can cause membrane rupture, which can result in ion loss and the release of cellular components (Tariq et al. 2019).

1.6.2.3 Viruses

Various significant oils have antiviral properties that are effective against a variety of RNA and DNA viruses, including herpes simplex Types 1 and 2 (HSV-1 and HSV-2), Type 2 influenza, Type 3 adenovirus, poliovirus, junin virus, and coxsackievirus B1 (Tariq et al. 2019).

1.6.3 Gene Modification of Integron

Gene modification of integrons can be an effective method for testing antimicrobial resistance of microorganisms that also shows multi-drug resistance. Integrons are genetic components of bacteria that can encourage the acquisition and expression of genes embedded within gene cassettes (GCs). The basic structure of an integron is given in Figure 1.5. Chromosomal integrons (CLs) and mobile integrons (MIs) are the two main categories of integrons that have been identified (Hall and Collis 1995). Numerous bacterial species have CLs on their chromosomes; an *in silico* investigation revealed that 17% of the sequenced bacterial genomes have such genetic arrangements. CLs are frequently found in bacteria from marine or terrestrial habitats, including *Xanthomonas* sp. and *Vibrio* sp. Owing to their ability to carry up to 200 cassettes, most of which encode proteins with unknown functions, CLs have also been referred to as "super integrons (Sls)". Additionally, cassettes without working reading frames may be carried by CLs. MIs are mobile genetic elements like transposons and plasmids that help spread them across bacteria. CLs are also known as "super integrons (Sls)" (Cambray, Guérout, and Mazel 2010) due to the 200 cassettes that they can contain, the majority of which encode proteins with uncertain functions. Furthermore, cassettes without active reading frames may be transported by CLs. MIs are mobile genetic elements like transposons and plasmids

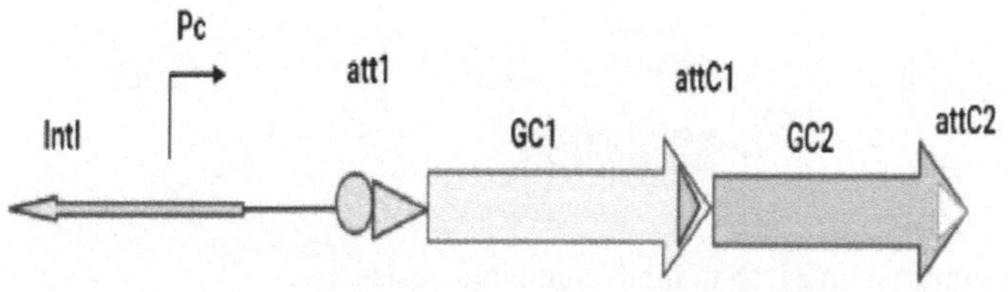

FIGURE 1.5 Basic structure of an integron.

that help spread them across bacteria. The MIs are often not self-transposable elements in and of themselves. MIs contain a small (10 GCs) number of GCs. The antibiotic-resistance determinants that are typically generated by the GCs mentioned so far in these MIs. This leads to the additional names "resistant integrons" (RIs) or "multi-drug resistance integrons" (MRIs) for MIs. Fewer Gram-positive bacteria than Gram-negative bacteria have been the subject of most descriptions of MIs. Five categories of MIs have been identified based on the intI protein's amino acid sequence. The three classes are the most frequently found. Class 4 and Class 5 integrons are rarely found and hence there are only limited information about both the integron classes.

1.6.4 Mechanism

An integron is defined as a functional platform (also known as a 5′ conserved segment, or 5′CS) that consists of three essential components: the intI gene, a distinct recombination site at attI (an attachment gene through which new genetic elements might enter), and the promoter Pc. The integrase protein intI, which is a member of the tyrosine recombinase family, is produced by the intI gene. By creating circular-shaped genetic molecules, tyrosine recombinases can keep cutting or adding genetic material. The GCs are non-replicative mobile elements that often connect an attC site with an open reading frame (ORF) that may encode genes for antibiotic resistance. An intI integrase-catalyzed site-specific recombination mechanism integrates or excises GCs from the functional platform. The annular GC3 gene cassette is inserted into the integron via the attI attachment site in the procedure described above. Additionally, the attC1 attachment site's GC1 gene cassette is eliminated. The variation in integron length demonstrates this. Excision and reintegration allow the cassettes to be moved around in their position (trial and error process). An integron carrying numerous antibiotic resistance genes that best combat the existing antimicrobial challenge is moved through this process to a site where it will be more strongly expressed. Figure 1.6 represents the mechanism of an integron to inhibit antimicrobial resistance (Dr. Jaydeep Ghosh).

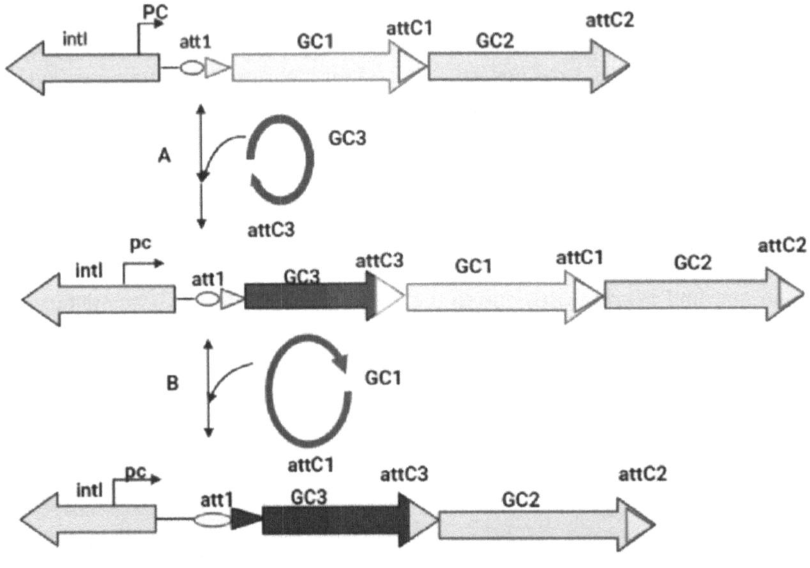

FIGURE 1.6 Mechanism of integron to inhibit antibiotic resistance.

1.7 CONCLUSION

It is evident that environmental contamination and antibiotic resistance both contribute to the spread of dangerous diseases among living organisms. Environmental contamination can, hence, be prevented if harmful anthropogenic activities are controlled. It is also necessary to understand that our activities are responsible for our and other biotic communities' survival. To prevent us from these diseases, we generate and use several antibiotics to destroy the pathogenic community. But during our subsequent usage, the pathogens become resistant to these antibiotics with the help of several defence mechanisms and integrons. So, methods like the usage of essential oils, gene modification of integrons, and inhibition of certain enzymes were discussed to prevent the microbes from acquiring antimicrobial resistance. Throughout this book chapter, we have suggested that environmental contamination and the generation of urban effluents are two inevitable processes. But we humans are responsible for other living species also. So, we are in a precarious position and we must reduce our harmful activities by promoting an eco-friendly system.

REFERENCES

"Contaminants in the Environment." https://oceanservice.noaa.gov/observations/contam/.

"Diseases Involving Sewage." 2022. *Environmental Public Health*. https://www.in.gov/health/eph/onsite-sewage-systems-program/diseases-involving-sewage/.

"Environmental Contaminants: Sources and Effects." http://www.Eurekaselect.Com. https://www.eurekaselect.com/chapter/13029.

Azevedo, Bruna Fernandes, Lorena Barros Furieri, Franck Maciel Peçanhã, Giulia Alessandra Wiggers, Paula Frizera Vassallo, Maylla Ronacher Simões, Jonaína Fiorim, et al. 2012. "Toxic effects of mercury on the cardiovascular and central nervous systems." *Journal of Biomedicine and Biotechnology*. doi:10.1155/2012/949048.

Burdett, Vickers. 1986. "Streptococcal tetracycline resistance mediated at the level of protein synthesis." *Journal of Bacteriology* 165, no. 2 (1 February): 564–569. doi:10.1128/jb.165.2.564-569.1986.

Bush, Karen, and Patricia A. Bradford. 2016. "B-Lactams and B-Lactamase inhibitors: An overview." *Cold Spring Harbor Perspectives in Medicine* 6, no. 8 (21 June): a025247. doi:10.1101/cshperspect.a025247.

Cambray, Guillaume, Anne-Marie Guérout, and Didier Mazel. 2010. "Integrons." *Annual Review of Genetics* 44, no. 1 (1 December): 141–166. doi:10.1146/annurev-genet-102209-163504.

Connell, Sean R., Catharine A. Trieber, George P. Dinos, Edda Einfeldt, Diane E. Taylor, and Knud H. Nierhaus. 2003. "Mechanism of Tet(O)-mediated tetracycline resistance." *EMBO Journal* 22, no. 4 (17 February): 945–953. doi:10.1093/emboj/cdg093.

Connell, Sean R., Dobryan M. Tracz, Knud H. Nierhaus, and Diane E. Taylor. 2003. "Ribosomal protection proteins and their mechanism ofTetracyclineResistance." *Antimicrobial Agents and Chemotherapy* 47, no. 12 (1 December): 3675–3681. doi:10.1128/aac.47.12.3675-3681.2003.

Davies, Julian, and Dorothy Davies. "Origins and evolution of antibiotic resistance." *Microbiology and Molecular Biology Reviews* 74, no. 3 (2010): 417–433.

Godt, Johannes, Franziska Scheidig, Christian Grosse-Siestrup, Vera Esche, Paul Brandenburg, Andrea Reich, and David A. Groneberg. "The toxicity of cadmium and resulting hazards for human health." *Journal of Occupational Medicine and Toxicology* 1 (2006): 1–6.

Grossman, Trudy H. "Tetracycline antibiotics and resistance." *Cold Spring Harbor Perspectives in Medicine* 6, no. 4 (2016): a025387.

Hall, Ruth M., and Christina M. Collis. 1995. "Mobile gene cassettes and integrons: Capture and spread of genes by site-specific recombination." *Molecular Microbiology* 15, no. 4 (1 February): 593–600. doi:10.1111/j.1365-2958.1995.tb02368.x.

Hooper, David C., and George A. Jacoby. 2015. "Mechanisms of drug resistance: Quinolone resistance." *Annals of the New York Academy of Sciences* 1354, no. 1 (17 July): 12–31. doi:10.1111/nyas.12830.

Integrons. PPT source- Dr. Jaydeep Ghosh, MSc PhD , Assistant Professor, St. Xavier's college.

Marilyn, C. Roberts, Patricia L. Keen, and Mark H. M. M. Montforts. 2011. Mechanisms of bacterial antibiotic resistance and lessons learned from environmental tetracycline-resistant bacteria. *Antimicrobial Resistance in the Environment.*

Roberts, Marilyn C. 2012. "Mechanisms of bacterial antibiotic resistance and lessons learned from environmental tetracycline-resistant bacteria." *Antimicrobial Resistance in the Environment*, 93–121.

Morais-Cabral, João H., Andrew P. Jackson, Clare V. Smith, Nita Shikotra, Anthony Maxwell, and Robert Liddington. 1997. "Crystal structure of the breakage–reunion domain of DNA gyrase." *Nature* 388, no. 6645 (28 August): 903–906. doi:10.1038/42294.

Munita, José M., and César A. Arias. 2016. "Mechanisms of antibiotic resistance." *Microbiology Spectrum*. doi:10.1128/microbiolspec.vmbf-0016-2015.

Rechenburg, Andrea, and Thomas Kistemann. 2009. "Sewage effluent as a source of *Campylobacter* sp. in a surface water catchment." *International Journal of Environmental Health Research (Print)*. doi:10.1080/09603120802460376.

Reygaert, Wanda C. "An overview of the antimicrobial resistance mechanisms of bacteria." *AIMS Microbiology* 4, no. 3 (2018): 482–501.

Rinkesh. 2024. "Conserve energy future - your source for green and sustainable [living, environmental news and information." *Conserve Energy Future*. https://www.conserve-energy-future.com/.

Solano-Gálvez, S.G., M.F. Valencia-Segrove, M.J.O. Prado, A.B.L. Boucieguez, D.A. Álvarez-Hernández, and R. Vázquez-López. (2020). Mechanisms of resistance to quinolones. *Antimicrobial Resistance-A One Health Perspective.*

Tariq, Saika, Saira Wani, Waseem Rasool, Khushboo Shafi, Muzzaffar A. Bhat, A. S. Prabhakar, Aabid H. Shalla, and Manzoor A. Rather. 2019. "A comprehensive review of the antibacterial, antifungal and antiviral potential of essential oils and their chemical constituents against drug-resistant microbial pathogens." *Microbial Pathogenesis* 134 (1 September): 103580. doi:10.1016/j.micpath.2019.103580,

van Grinsven, H.J., Ward, M.H., Benjamin, N. *et al.* "Does the evidence about health risks associated with nitrate ingestion warrant an increase of the nitrate standard for drinking water?." *Environ Health* 5 (2006): 26. doi: 10.1186/1476-069X-5-26

Worthington, Roberta J., and Christian Melander. 2013. "Overcoming resistance to B-Lactam antibiotics." *Journal of Organic Chemistry* 78, no. 9 (28 March): 4207–4213. doi:10.1021/jo400236f.

2 Antimicrobial Resistance in Aquaculture at the Juncture of Global Warming and Other Effects

K.R. Preethy, M. Chamundeeswari, G. Sivaranjani, and K. Aishwarya

2.1 INTRODUCTION

Feeding a rapidly expanding human population while managing a deteriorating environment caused by food production will be a major problem in the years to come (UN, 2018 This is especially true for low- and middle-income countries (LMICs), where there is a likelihood that the need for animal protein will increase (Robinson & Pozzi, 2011) and where recent environmental changes, such as severe droughts, floods, and wildfires, have resulted in severe food crises (Suweis et al., 2015; FAO, 2017). As a result, the 2030 UN Agenda for Sustainable Development Goals, which aims to eradicate poverty and prevent environmental degradation, places a high priority on food security (UN, 2018). In order to meet these objectives, food production must be raised, but more importantly, proper husbandry practices must be followed to minimize its detrimental effects on the environment. The simple response to today's rising food demand is intensification of production, which leads to risks to the environment and human health by increasing water usage or excessive antibiotic use (Herrero et al., 2015; Van Boeckel et al., 2015; Schar et al., 2018). According to various studies (Springmann et al., 2016; Thilsted et al., 2016; Froehlich et al., 2018), changing the human diet to include more fish and seafood might help to meet the requirement for protein that would support both human and environmental health. In fact, the aquaculture industry is expected to increase by 62% over the same time period, supporting a 27% rise in seafood and fish consumption by 2030 (Fish to 2030: Prospects for Fisheries and Aquaculture, 2013). Over 100 million individuals are projected to depend on aquaculture for a living, making it a substantial contributor to the income of many households (FAO, 2018).

With emerging infectious diseases (EIDs) predicted to rise with rising temperatures (Marcos-López et al., 2010; Karvonen et al., 2010; Vezzulli et al., 2016; Bondad-Reantaso, 2016; FAO, 2018), fish farming primarily relies on the use of antibiotics to treat infectious illnesses that endanger output. Acute hepatopancreatic necrosis illness, streptococcal infections, and *edwardsiellosis* outbreaks, for instance, are frequently seen when the temperature rises (Mohanty & Sahoo, 2007; Kayansamruaj et al., 2014; Bondad-Reantaso, 2016). Given this situation, it is anticipated that antimicrobial usage will increase over the next several years, particularly in LMICs, as production methods become more intensive in order to satisfy consumer demand for animal products and economic needs (Van Boeckel et al., 2015; Schar et al., 2018). However, the combined use of antimicrobial medications in aquaculture and land-derived pollution into watercourses contribute to the selection, development, and spread of drug-resistant bacteria, posing a significant hazard to

DOI: 10.1201/9781003268925-2

public health (Van Boeckel et al., 2015; Schar et al., 2018; Cabello et al., 2016). Rising temperatures have been linked to greater rates of antibiotic resistance in terrestrial bacteria, presenting a gloomy prognosis in the context of global climate warming (MacFadden et al., 2018). Recent research has also indicated that antimicrobial usage may not be the primary driver influencing the selection and establishment of AMR.

2.2 ANTIMICROBIALS USED IN AQUACULTURE AND THEIR MECHANISMS OF RESISTANCE

Among the most frequently used antimicrobial substances in aquaculture, quinolones (such as oxolinic acid, flumequine, and enrofloxacin), tetracyclines (such as oxytetracycline (OTC)), and phenicols (such as florfenicol) have been widely used to control bacterial fish diseases in salmon farming. Despite currently only making up less than 1% of global antimicrobial usage, quinolones are nevertheless being increasingly utilized in the development of aquaculture in numerous Asian nations. According to reports, quinolones, tetracyclines, and phenicols are selective for a range of antimicrobial resistance genes (ARGs) and frequently appear in mobile genetic elements that encourage their spread (such as transposons, plasmids, and integrons) (Akinbowale et al., 2007).

2.2.1 Quinolone Resistance

DNA gyrase and topoisomerase IV are two of the bacterial enzymes that quinolones target. The primary function of the tetrameric enzyme DNA gyrase, which is encoded by the genes gyrA and gyrB, is to catalyze the negative supercoiling of bacterial DNA. The parC and parE genes produce topoisomerase IV, a tetrameric enzyme that decatenates and relaxes the activity of daughter replicons after DNA replication (Anderson et al., 1998; Hawkey, 2003). Topoisomerase gene mutations (gyrA, gyrB, parC, and parE) and mutations that decrease drug accumulation by boosting absorption or lowering efflux are the main causes of the development of quinolone resistance (Drlica & Zhao, 1997; Ruiz, 2003).

2.2.2 Tetracycline Resistance

Widely used in fish farming, oxytetracycline is a broad-spectrum bacteriostatic antibiotic.

that is effective against a variety of Gram-positive and Gram-negative bacteria. Prokaryotic tetracyclines bind reversibly to the 70S ribosome and prevent the production of proteins (Chopra, 1985; Roberts, 1996). Active efflux, ribosomal protection, ribosomal RNA alterations, and tetracycline inactivation are some of the tetracycline resistance mechanisms (Speer & Salyers, 1989; Salyers et al., 1990; Burdett, 1991; Levy, 1992; Speer et al., 1992; Taylor & Chau, 1996). It has been discovered that one or more members of the Tet family of proton-dependent efflux pumps or ribosomal protection by cytoplasmic proteins found frequently in Gram-negative bacteria plays a significant role in the tetracycline resistance of bacteria linked to fish farms (Roberts, 2005; 2012).

2.2.3 Phenicol Resistance

Florfenicol, a synthetic fluorinated derivative of chloramphenicol, has a bacteriostatic effect based on a reversible binding to the 50S subunit of 70S bacterial ribosomes, which hinders peptide elongation (Schwarz et al., 2004). A frequent route of bacterial resistance to chloramphenicol involves chloramphenicol acetyltransferases (CATs), which inactivate florfenicol by replacing a hydroxyl group with a fluorine atom (Shaw & Leslie, 1991; Schwarz et al., 2004). For the fish farming business, florfenicol is a very valuable medicine due to its potency against a variety of important fish diseases (Fukui et al., 1987). RNA methyltransferases, specialized hydrolases, and specific and non-specific drug transporters are among the resistance mechanisms of florfenicol (Schwarz et al., 2004).

2.3 RESISTANCE AGAINST ANTIMICROBIALS

The capacity of microorganisms to thrive and endure the effects of antimicrobials designed to suppress or kill them is known as antimicrobial resistance. Every time antimicrobials are administered to people, animals, or plants, there is a threat that microorganisms may develop the ability to resist treatments, eventually becoming less susceptible to the action of the drug. As a result, a rise in infections and production losses in the case of agriculture is observed.

2.3.1 AMR in Aquaculture

Bacteria that are Gram-negative like *Aeromonas hydrophila*, *Vibrio harveyi*, *Vibrio anguillarum*, *Aeromonas salmonicida*, *Edwardsiella tarda*, *Flavobacterium psychrophilum*, *Citrobacter freundii*, *Yersinia ruckeri*, and *Pseudomonas fluorescens* are primarily responsible for the most frequently observed infections in aquaculture. It is also caused by *Mycobacterium* sp. (Lewbart, 2001). *Aeromonas hydrophila* has been recognized as the most prevalent bacterial pathogen in freshwater aquaculture (Igbinosa et al., 2012).

To control such infections, antimicrobials are intensively used. The prolonged use of the drugs leads to development of AMR in these infectious microorganisms. Due to both extensive anthropogenic activity (Levy & Marshall, 2004; Davies & Davies, 2010; Kolter & Van Wezel, 2016) as well as natural processes (D'Costa et al., 2011), bacteria are developing resistance to a wide variety of antibiotics. Antimicrobial resistance (AMR) in bacteria can arise spontaneously, via transformation, transduction, or by conjugation (Chamier et al., 1993; Wright, 2007; Iwasaki & Takagi, 2009; Aminov, 2011; Taylor et al., 2011; Stalder et al., 2012; Marti et al., 2014). They can also acquire AMR through horizontal gene transfer (HGT). Due to the microbial community's genetic flexibility, resistance genes can spread among various environmental bacterial populations and communities.

2.3.2 Emanation of Multi-Antibiotic Resistance in Catfish Aquaculture

More than 50% of the fish eaten globally is from aquaculture, which endows protein to many countries in Latin America, Asia, Africa, etc. Of which the catfish species are most commonly cultivated. Antimicrobials are used in the farming of catfish as curative and preventative agents. Oxytetracycline – a blend of sulfadimethoxine, florfenicol, and ormetoprim – is given to catfish in the United States as it is the only drug approved by the FDA for the species (FDA, 2016; FDA, 2016). The use of drugs such as flumequine, oxytetracycline, and trimethoprim along with their combinations is only allowed for catfish in Italy (Labella et al., 2013). Such drugs, when put to wrong use, lead to the rise of antimicrobial resistance. Data on the exact dosage of antimicrobials in aquaculture is meagre and hence exact measurements are not mentioned here. The possible pathways of antibiotics and the dispersion of multi-antimicrobial resistance are explained in Figure 2.1.

The catfish species is found to have resistance to multiple drugs such as tetracycline, ampicillin, kanamycin, and oxytetracycline. The development, rise, and propogation of such multi-antimicrobial-resistant species are not restricted only to microbes from catfish; rather, resistance has also been found in surrounding species other than those from aquaculture (McPhearson et al., 1991).

2.3.3 Transfer of Antibiotic Resistance between Bacteria

Bacteria as a class or species are not always equally vulnerable to or resistant to a given antibiotic. Within bacterial groups that are closely related, resistance levels might vary substantially. The least

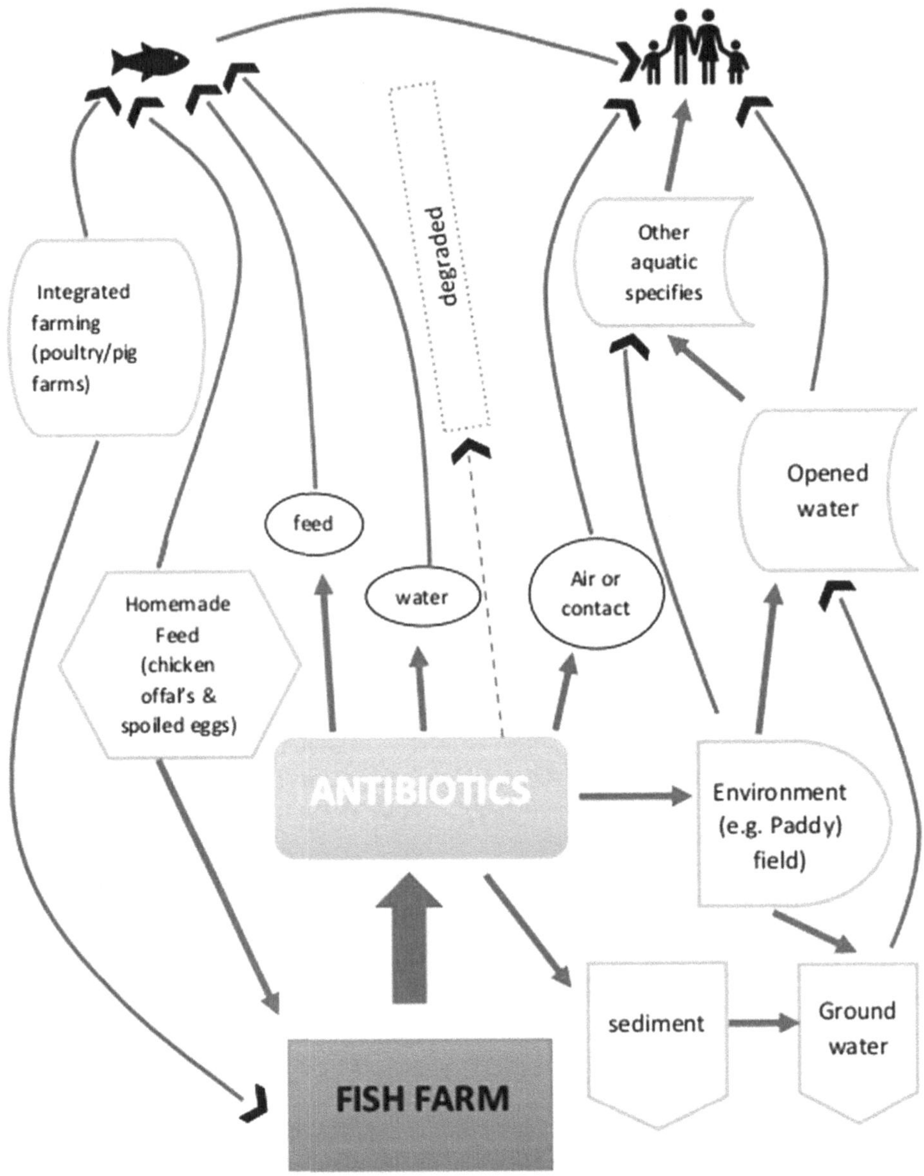

FIGURE 2.1 Flowchart of antibiotics and the dispersion of multi-antimicrobial resistance.

drug concentration necessary to stop bacterial growth is known as the minimum inhibitory concentration, or MIC, and it is used to determine both susceptibility and resistance. The degree of resistance will vary depending on the inherited genes and the bacterial species, as bacteria may also acquire resistant genes from other related organisms (Coculescu, 2009; Martinez, 2014). There was natural resistance before antibiotics were used. Such resistance does not necessitate an evolutionary connection between the bacterial species and the antibiotic. This type of resistance is unrelated to the use of antibiotics and is brought on by the structural features of bacteria. For example, vancomycin does not pass through the outer membrane of Gram-negative bacteria; hence this type of bacterium naturally resists vancomycin (Cesur & Demiroz, 2013).

2.3.4 Transfer Mechanism of Resistance between Bacteria

A mobilome is a collection of movable genetic components found in a genome that may travel across aquatic microorganisms. An important phenomenon involved in the development of aquatic pathogens that are resistant to antibiotics is horizontal gene transfer. In addition to the abundance of viruses and other gene transfer vectors in seawater, biofilm communities of aquatic bacteria attached to sediment clays, organic particle matter, and sands, as well as fish farm structures, aid in the development of antibiotic resistance (Abe et al., 2020). DNA produced by bacteriophage lysis and plasmids can both function as carriers for horizontal gene transfer. Antimicrobial resistance genes may be present in plasmid and bacteriophage bare DNA and may be expressed once they infect bacteria. Aquatic bacteria, such as *Vibrio* sp., are inherently capable of absorbing DNA, enabling transformation to occur in the aquatic environment (Cabello et al., 2013) (Figure 2.2).

2.3.5 Effects of Antimicrobial-Resistant Bacteria Transport from Aquaculture on Human Population

Antimicrobial resistance in bacteria that cause infections in humans has three negative effects: (1) more severe infections (2) more treatment failures and (3) Increased Risk of Disability (Kruse & Sørum, 1994).

Increase in infections: Antimicrobials may alter the microbiota in the human intestinal system, putting those who have been treated at higher risk for specific illnesses. Due to bacteria that are resistant to the antimicrobial agent, people who are taking an antibiotic for any reason are

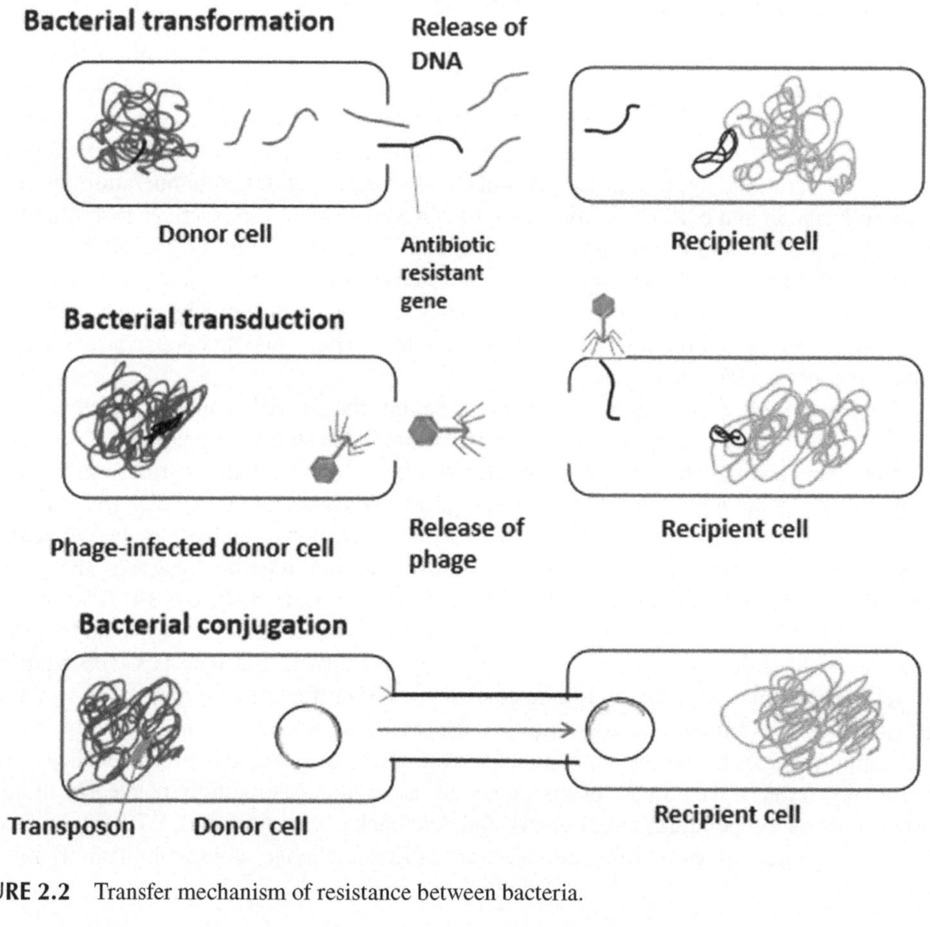

FIGURE 2.2 Transfer mechanism of resistance between bacteria.

more likely to get an illness. People who are exposed to antimicrobial agents for unrelated reasons, like the treatment of an upper respiratory tract infection, are at increased risk for infection due to antimicrobial-resistant *Salmonella* species, as shown in case–control studies involving people infected with *Salmonella* species that are drug resistant (Barza & Travers, 2002). This elevated risk can be expressed using an attributable fraction, such as the percentage of *Salmonella* infections that occurred as a consequence of the *Salmonella* species being resistant to the antibiotic therapy (i.e. infections occur as a consequence of the person taking the antimicrobial agent for an unrelated reason). Even though there is currently no data from aquaculture-related studies, it is reasonable to speculate that the same trend that has been shown to occur for *Salmonella* species can occur with other drug-resistant pathogenic strains for which resistance may have originated in aquaculture, and that antibiotic therapies may put the patient at risk for infection due to such drug-resistant pathogens.

Increased infection severity and failure to treatments: Antimicrobial resistance's higher frequency of treatment failure and increased severity of infection may lead to increased bloodstream infection frequency, increased hospitalization, or increased death (Kruse & Sørum, 1994). Case-control studies of fluoroquinolone-resistant *Campylobacter* have shown prolonged disease duration (Smith et al., 1999; Neimann et al., 2003), while investigations of infections caused by quinolone-resistant *Salmonella typhimurium* have shown higher infection severity (Helms et al., 2004). Increased morbidity or mortality has been shown for antimicrobial-resistant non-*Typhi Salmonella* serotypes and *Campylobacter* (Molbak, 2005). It is logical to expect that other drug-resistant human diseases, for which resistance may have its roots in aquaculture, will exhibit the same behaviour that has been observed for *Salmonella* and *Campylobacter* species.

2.3.6 Global Warming and Antibiotic Resistance

The survival of bacteria is influenced by a number of factors, including temperature. Therefore, it is crucial to comprehend how temperature fluctuations affect the evolution of antibiotic resistance. Since the origin of life, temperature fluctuations have always existed, and organisms have developed a variety of defence systems to withstand them and continue to survive. Numerous research studies and observations imply that the physiological stress reactions to temperature by bacteria may be implicated in and partially co-opted for the physiological stress reactions to antibiotics and the development of antibiotic resistance. By changing the cellular equilibrium, antibiotics affect particular bacterial targets. Analyzing the impact of temperature on bacterial cells reveals similar alterations. It is interesting how antibiotic-induced changes in bacterial cells match up with temperature rises. Because antibiotics and heat have similar mechanisms of action, a rise in antibiotic resistance may occur (McGough et al., 2020).

The danger of AMR will likely be highest in the nations most at risk from climate change, harming human health in ways that go beyond the aquaculture industry and emphasizing the urgency of taking immediate action. The previous three decades have each been the warmest on record as a result of anthropogenic emissions of greenhouse gases, which pushed post-industrial warming of the climate (Pachauri et al., 2014), and as of 2017, the average global temperature had increased by over 1°C from pre-industrial times (Allen et al., 2018). Without mitigating measures, the estimated increase in global temperature by the end of the twenty-first century is up to 4.8°C (Pachauri et al., 2014). By 2040, if things continue as they are, we may have exceeded the 1.5°C global temperature increase threshold (Allen et al., 2018). The Arctic may warm by as much as 11°C (Pachauri et al., 2014), yet the rates and impacts of climate change are not uniform everywhere. Due to the complexity and diversity of microbial communities, the spread of infectious illnesses in environments with changing climates cannot be automatically predicted with absolute confidence. However, it is generally known that a wide range of infectious disease pathogens multiply noticeably more rapidly when ambient temperatures rise. Lemery and Auerbach (Lemery et al., 2017) stated that while it is possible that new illnesses won't emerge due to climate change, it will significantly raise the danger to human populations. As a result, the need for efficient vaccinations and preventative advice will become more essential. Relying solely on the development of novel antibiotics is no longer an option due to the pervasive and growing threat posed by drug resistance, the significant difficulties associated with identifying and obtaining regulatory approval for novel classes of antimicrobial agents, the notable capacity of pathogenic bacteria to build resistance to antibiotics, and the fact that global warming is a "threat multiplier" for the spread of infectious diseases and antibiotic resistance (Fouladkhah et al., 2020). By changing the cellular equilibrium, antibiotics affect particular bacterial targets. Analyzing the impact of climate on bacterial cells reveals similar alterations. It is interesting how antibiotic-induced changes in bacterial cells match up with temperature rises. Antibiotic resistance may continue to rise as a result of the similar actions of both temperature and antibiotics (McGough et al., 2020). Aminoglycosides are a family of antibiotics that harm the structure of cells by attaching to the ribosomes and causing an increase in misfolded proteins in cells. Qualitatively comparable alterations may also be brought on by temperature rises and the strains of heat shock (Greulich et al., 2015; Ramirez & Tolmasky, 2010). Gentamicin and tobramycin, two aminoglycosides, respectively stimulate heat shock genes in *Bacillus subtilis* and *P. aeruginosa* (Lin et al., 2005; Kindrachuk et al., 2011).

2.3.6.1 Effect Cross-Protection

Collateral resistance, also known as cross-protection, is the phenomenon wherein mutations that provide resistance to thermal stress can also confer drug resistance and vice versa (Rodriguez-Verdugo et al., 2013). The consequences of collateral resistance or cross-protection are based on the idea that bacterial communities have extremely flexible genotypes that enable them to adapt to shifting environmental conditions (Dragosits et al., 2013). Cross-protection is the term used to describe acquired resistance to a particular stressor following prior exposure to another stressor agent. In a cross-protection experiment, *E. coli* cells were exposed to high temperature for 2,000 generations (Rodriguez-Verdugo et al., 2013). Even though the same antibiotic was not used during the experiment, 10% of the experimental lines eventually developed rifampicin resistance due to *de novo* mutations after hundreds of generations. In this instance, a temperature of 42°C served as the reference of the selective pressure. Because both rifampicin and temperature operate on the same target of selection, the RNA polymerase, which is responsible for controlling gene expression and the transcription of DNA into RNA, the mechanism of cross-resistance developed. Gene expression under conditions of extreme temperature stress changed as a result of mutations in the RNA polymerase active site (Rodriguez-Verdugo et al., 2016).

2.3.6.2 Modifications Induced by Temperature

The microbial community is then kept in check as a result of temperature influences on antibiotic resistance and associated changes. A seawater bacterial strain of the genus *Roseovarius* that grew at 33°C instead of 25°C, which is the ideal temperature, and gave rise to a biofilm is an example of how structural alteration may be facilitated by the elevated temperature. The development of biofilms is caused by an adaptation to high temperatures, which results in an increase in biofilm development at the air–liquid interface. Thus, as with biofilms, temperature rises alter physiological traits that favour particular adaptations. These modifications could increase antibiotic tolerance, which would have wider ramifications for marine bacteria that are encountering greater seawater temperatures as a result of climate change (Kent et al., 2018). The temperature promotes the early emergence of antibiotic resistance mutations. According to the idea of evolvability (Payne & Wagner, 2019), the temperature can actually promote adaptive evolution by enhancing phenotypic and genotypic variety in a population.

2.4 SUBSTITUTION OF PHYTOCOMPOUNDS AS A DIFFERENT ANTIMICROBIAL APPROACH IN AQUACULTURE

Regardless of investing in developing the aquaculture sector, farmers are affected due to the upsurge of contagious diseases caused by pathogens. To curb these, antimicrobials are generally used, but after a while injudicious use of them turns out to be a real problem, causing resistance in pathogens. Currently, natural compounds called phytocompounds can be used as a substitute for antimicrobials, which as a whole do not cause any harm to the environment and the lives of species as well. Some classes of phytocompounds are discussed below.

2.4.1 Flavanoids

Plants naturally produce flavonoids, which are a wide group of polyphenolic chemicals with a structural benzo-pyrone. They are produced through the phenylpropanoid pathway (Kumar & Pandey, 2013). In the plant kingdom, over 9,000 distinct flavonoids have reportedly been found to date (Markham & Andersen, 2006), and at least a hundred of these are edible components (Tuli, 2019). Flavonoids are classified into flavanones, flavonols, flavones, isoflavones, and anthocyanins based on their chemical makeup. One study claims that the flavan backbone of flavonoids is made up of two benzene rings A and B joined by a heterocyclic pyrone ring C (Figure 2.3) (Panche, 2016). Having a carbonyl group at the beginning of a compound allows it to be classified as a flavonoid compound.

2.4.2 Alkaloids

A group of naturally occurring organic bases that contain nitrogen is known as alkaloids. Alkaloids have a wide range of physiological effects on humans and other animals (The Editors of Encyclopaedia Britannica, 2020). They have been detected in various microbes, mammals, insects, and marine invertebrates in addition to being isolated from plants (Bribi, 2018). Sparteine, morphine, quinine, ephedrine, and nicotine are a few examples of alkaloids. Over 40,000 different alkaloid compounds have reportedly been discovered (Buckingham et al., 2010).

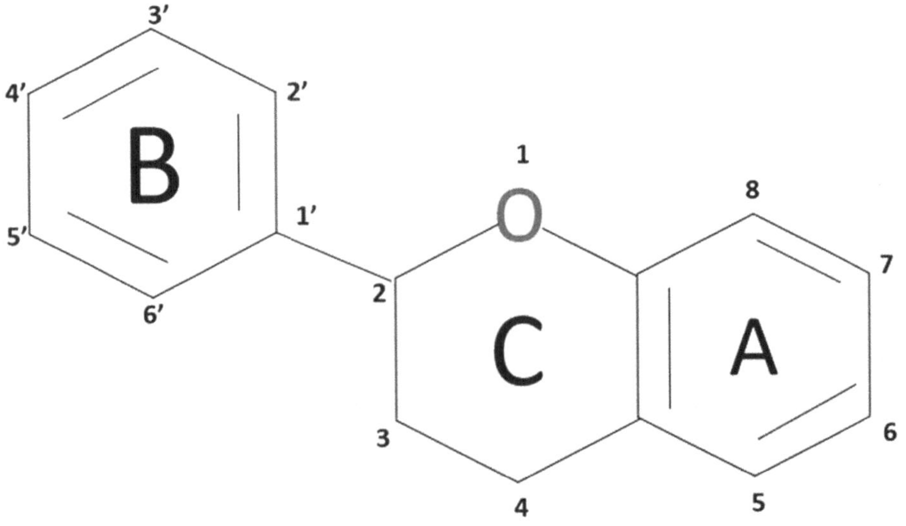

FIGURE 2.3 Flavonoid structure.

2.4.3 Phenolic Acids

One of the primary families of plant phenolic compounds is phenolic acids, which are defined as phenolic compounds with one carboxylic acid group. Free phenolic acids, including phenylacetic, benzoic, and cinnamic acids, have superior water solubility and bioavailability compared to flavonoids. A particular family of phenolic acids is implicated in defence mechanisms against biotic and abiotic stressors (Dicko et al., 2005). Phenolic acids have been demonstrated to reduce oxidative deterioration when added as attributes in dispersion model systems (Lisete-Torres et al., 2012). They have significant lipid and water solubility. Flavonoids cannot be absorbed in the stomach, and only a limited amount is carried through the intestinal wall into the circulation (Chen et al., 2020). Carnosic acid, gallic acid, rosmarinic acid, and caffeic acid are a few examples of phenolic acids.

2.4.4 Terpenoids

Terpenoids, commonly referred to as isoprenoids or terpenes, are a broad class of natural substances that are present in almost all living things (Oldfield & Lin, 2012). One of the largest classes of known natural compounds is the terpenoid family, which has around 60,000 recognized structural variants (Berthelot et al., 2021). Compared to other living things, plants have a lot of terpenoids. According to one study, there could be tens of thousands of different terpenoid molecules in plants, which include both terpenes and compounds with terpene moieties coupled to other moieties obtained from other pathways (Chen et al., 2011). According to a recent study, numerous plants that contain terpenoids have been extensively employed in traditional medicine due to their potent anti-inflammatory and pain-relieving effects (Agatonovic-Kustrin & Morton, 2018). It has usually been hypothesized that the lineage-specific terpenoids that have evolved throughout the evolution of green plants have a role in the ecological interactions of plants with biotic and abiotic elements of their environment (Pichersky & Raguso, 2016). These functions include disease and herbivore protection, as well as signals and incentives for helpful organisms like pollinators and mycorrhiza (Heiling et al., 2010).

2.5 CONCLUSION

Aquaculture has undeniable benefits in terms of the economy and society, but it also has some serious problems, one of which is antibiotic resistance. The use of antibiotics in animals has been linked to the establishment of resistant strains of bacteria and the spread of resistance to human pathogens. Additionally, when temperatures rise and the climate changes, bacteria adapt and a large rise in antibiotic resistance results. In conclusion, microorganisms that encounter increases and decreases in temperature either within a host or in the external environment are likely to develop a thermal adaptation. Increased antibiotic resistance is one of the possible side effects of these thermal adaptations. Even though the majority of microorganisms are not pathogens, the larger community of microbes acts as a reservoir for genes that make diseases resistant to antibiotics. Aquaculture may eventually switch to phytocompounds as an antibacterial agent. A wide variety of potential actions in phytocompounds, ranging from moderate to high levels of antibacterial activity toward freshwater and marine diseases, have been demonstrated through cutting-edge research and emerging technologies. The use of these phytocompounds in aquaculture would unquestionably be a step in the right direction toward the utilization of therapeutic plants and herbs.

REFERENCES

Abe K, Nomura N, Suzuki S. 2020. Biofilms: Hot spots of horizontal gene transfer (HGT) in aquatic environments, with a focus on a new HGT mechanism. *FEMS Microbial Ecol.* 96. doi: 10.1093/femsec/fiaa031

Agatonovic-Kustrin S, Morton DW. 2018. *The Cosmeceutical Properties of Compounds Derived from Marine Algae. Marine Macro-Microalgae: Overview.* CRC Press, Taylor and Francis, 198.
Akinbowale OL, Peng H, Barton MD. 2007. Diversity of tetracycline resistance genes in bacteria from aquaculture sources in Australia. 103(5). doi: 10.1111/j.1365-2672.2007.03445.x
Allen MR, Dube OP, Solecki W, et al. 2018. *Global warming of 1.5 °C: An IPCC Special Report on the Impacts of Global Warming of 1.5 °C above Pre-Industrial Levels and Related Global Greenhouse Gas Emission Pathways, in the Context of Strengthening the Global Response to the Threat of Climate Change, Sustainable Development, and Efforts to Eradicate Poverty.* Intergovernmental Panel on Climate Change. Geneva, Switzerland. https://www.ipcc.ch/site/assets/uploads/sites/2/2019/06/SR15_Full_Report_Low_Res.pdf
Anderson VE, Gootz TD., & Osheroff N. (1998). Topoisomerase IV catalysis and the mechanism of quinolone action. *Journal of Biological Chemistry, 273*(28), 17879–17885.
Aminov RI. 2011. Horizontal gene exchange in environmental microbiota. *Front Microbiol.* 2(158). https://doi.org/10.3389/fmicb.2011.00158
Barza M, Travers K. 2002. Excess infections due to antimicrobial resistance: The "Attributable fraction". *Clin Infect Dis.* 34(3):126–130. https://doi.org/10.1086/340250
Berthelot K, Estevez Y, Deffieux A, Peruch F. 2012. Isopentenyl diphosphate isomerase: A checkpoint to isoprenoid biosynthesis. *Biochimie.* 94(8):1621–1634. doi: 10.1016/j.biochi.2012.03.021
Bontad-Reantaso MG 2016. Acute hepatopancreatic necrosis disease (AHND) of penaeid shrimps: Global perspective. In R. V. Pakingking Jr, E. G. T. de Jesus-Ayson, & B. O. Acosta (Eds.), *Addressing Acute Hepatopancreatic Necrosis Disease (AHPND) and Other Transboundary Diseases for Improved Aquatic Animal Health in Southeast Asia. Aquaculture Department, Southeast Asian Fisheries Development Center, Tigbauan, Iloilo, Philippines.*
Bribi N. 2018. Pharmacological activity of alkaloids: A review. *Asian J Botany.* 1:1–6 doi:10.63019/ajb.v1i2.467
Buckingham J, Baggaley KH, Roberts AD, Szabo LF. 2010. *Dictionary of Alkaloids with CD-ROM.* CRC Press, Boca Raton, FL. https://doi.org/10.1201/EBK1420077698
Burdett V. (1991). Purification and characterization of Tet (M), a protein that renders ribosomes resistant to tetracycline. *Journal of Biological Chemistry, 266*(5), 2872–2877.
Cabello FC, Godfrey HP, Buschmann AR, et al. 2016. Aquaculture as yet another environmental gateway to the development and globalisation of antimicrobial resistance. 16(7) doi: 10.1016/S1473-3099(16)00100-6
Cabello FC, Godfrey HP, Tomova A, et al. 2013. Antimicrobial use in aquaculture re-examined: Its relevance to antimicrobial resistance and to animal and human health. *Environ Microbiol.* 15(7). doi: 10.1111/1462-2920.12134.
Cesur S, Demiroz AP. 2013. Antibiotics and the mechanisms of resistance to antibiotics. *Med J Islamic World Acad Sci.* 21(4):138–142.
Chamier B, Lorenz MG, Wackernagel W. 1993. Natural transformation of *Acinetobacter calcoaceticus* by plasmid DNA adsorbed on sand and groundwater aquifer material. *Appl Environ Microbiol.* 59:1662–1667. doi: 10.1128/aem.59.5.1662-1667.1993
Chen F, Tholl D, Bohlmann J, Pichersky E. 2011. The family of terpene synthases in plants: A mid-size family of genes for specialized metabolism that is highly diversified throughout the kingdom. *Plant J.* 66(1):212–229. doi: 10.1111/j.1365-313X.2011.04520.x
Chen J, Yang J, Ma L, Li J, Shahzad N, Kim CK. 2020. Structure-antioxidant activity relationship of methoxy, phenolic hydroxyl, and carboxylic acid groups of phenolic acids. *Scientific Rep.* 10:2611. Doi: https://doi.org/10.1038/s41598-020-59451-z
Chopra I. (1986). Genetic and biochemical basis of tetracycline resistance. *Journal of Antimicrobial Chemotherapy, 18*(Supplement_C), 51–56.
Coculescu BI. 2009. Antimicrobial resistance induced by genetic changes. *J Med Life.* 2(2):114–123.
D'Costa VM, King CE, Kalan L, et al. 2011. Antibiotic resistance is ancient. *Nature* 477:457–461
Davies J, Davies D. 2010. Origins and evolution of antibiotic resistance. 74(3):417–433. doi: 10.1128/MMBR.00016-10
Dicko MH, Gruppen H, Traoré AS, et al. 2005. Evaluation of the effect of germination on phenolic compounds and antioxidant activities in sorghum varieties. *J Agric Food Chem.* 53:2581–2588. doi: 10.1021/jf0501847
Dragosits M, Mozhayskiy V, Quinones-Soto S, et al. 2013. Evolutionary potential, cross-stress behavior and the genetic basis of acquired stress resistance in *Escherichia coli. Mol. Syst. Biol.* 9:643. doi: 10.1038/msb.2012.76.
Drlica K, Zhao X. (1997). DNA gyrase, topoisomerase IV, and the 4-quinolones. *Microbiology and Molecular Biology Reviews, 61*(3), 377–392.

Encyclopedia Britannica. 2020. *The Editors of Encyclopaedia*. Alkaloid. https://www.britannica.com/science/alkaloid
FAO. 2017. *The Future of Food and Agriculture – Trends and Challenges* (Rome).
FAO. 2018. *Impacts on Climate Change on Fisheries and Aquaculture – Synthesis on Current Knowledge, Adaptation and Mitigation Options* (Rome).
FAO. 2018. *The State of World Fisheries and Aquaculture 2018 – Meeting the Sustainable Development Goals* (Rome).
FDA. 2016. NADA number: 038–439. http://www.accessdata.fda.gov/scripts/animaldrugsatfda/details.cfm?dn=038-439.
FDA. 2016. NADA number: 141–246. http://www.accessdata.fda.gov/scripts/animaldrugsatfda/details.cfm?dn=141-246.
Froehlich HE, Runge CA, Gentry RR. et al. 2018. Comparative terrestrial feed and land use of an aquaculture-dominant world. 115(20). https://doi.org/10.1073/pnas.1801692115
Fouladkhah AC, Thompson B, Camp JS. 2020. The threat of antibiotic resistance in changing climate. *Microorganisms*. 8(5):748. doi: 10.3390/microorganisms8050748
Fukui H, Fujihara Y, Kano T. (1987). In vitro and in vivo antibacterial activities of florfenicol, a new fluorinated analog of thiamphenicol, against fish pathogens. *Fish Pathology, 22(4)*, 201–207.
Greulich P, Scott M, Evans MR, Allen RJ. 2015. Growth-dependent bacterial susceptibility to ribosome-targeting antibiotics. *Mol Syst Biol*. 11(3):796. doi: 10.15252/msb.20145949.
Hawkey PM. (2003). Mechanisms of quinolone action and microbial response. *Journal of Antimicrobial Chemotherapy*, *51*(suppl_1), 29–35.
Heiling S, Schuman MC, Schoettner M, et al. 2010. Jasmonate and ppHsystemin regulate key malonylation steps in the biosynthesis of 17-hydroxygeranyllinalool diterpene glycosides, an abundant and effective direct defense against herbivores in *Nicotiana attenuata*. *Plant Cell*. 22(1):273–292. doi: 10.1105/tpc.109.071449
Helms M, Simonsen J, Molbak K. 2004. Quinolone resistance is associated with increased risk of invasive illness or death during infection with Salmonella serotype Typhimurium. *J Infect Dis*. 190(9):1652–1604. https://doi.org/10.1086/424570
Herrero M, et al. 2015. Livestock and the environment: What have we learned in the past decade? *Annu Rev Environ Resour*. 40:177–202. https://doi.org/10.1146/annurev-environ-031113-093503
Igbinosa EO, Odjadjare EE, Igbinosa IH, Orhue PO, Omoigberale MNO, Amhanre N I, et al. (2012). Antibiotic synergy interaction against multidrug-resistant Pseudomonas aeruginosa isolated from an abattoir effluent environment. The Scientific World Journal, Article ID 308034, 5 pages
Iwasaki W, Takagi T. 2009. Rapid pathway evolution facilitated by horizontal gene transfers across prokaryotic lineages. *PLoS Genet*. https://doi.org/10.1371/journal.pgen.
Karvonen A, Rintamäki P, Jokela J, et al. 2010. Increasing water temperature and disease risks in aquatic systems: climate change increases the risk of some, but not all, diseases. *Int J Parasitol*. 40(13):1483–1488. https://doi.org/10.1016/j.ijpara.2010.04.015
Kayansamruaj P, Pirarat N, Hirono et al. 2014. Increasing of temperature induces pathogenicity of *Streptococcus agalactiae* and the upregulation of inflammatory related genes in infected *Nile tilapia* (*Oreochromis niloticus*). 172(1–2). doi:10.1016/j.vetmic.2014.04.013
Kent AG, Garcia CA, Martiny AC. 2018. Increased biofilm formation due to high temperature adaptation in marine *Roseobacter*. *Nat Microbiol*. 3:989–995. doi: 10.1038/s41564-018-0213-8
Kindrachuk KN, Fernández L, Bains M, Hancock REW. 2011. Involvement of an ATP-dependent protease, PA0779/AsrA, in inducing heat shock in response to tobramycin in *Pseudomonas aeruginosa*. *Antimicrob Agents Chemother*. 55(5):1874–1882. doi: 10.1128/AAC.00935-10
Kolter R, van Wezel GP. 2016. Goodbye to brute force in antibiotic discovery? 27(1). doi: 10.1038/nmicrobiol.2015.20
Kruse H, Sørum H. 1994. Transfer of multiresistance plasmids between bacteria of diverse origin in natural micro-environments. *Appl Environ Microbiol*. 60(11):4015–4021. doi: 10.1128/aem.60.11.4015-4021.1994
Kumar S, Pandey AK. 2013 Chemistry and biological activities of flavonoids: An overview. *Sci World J*. doi: 10.1155/2013/162750
Labella A, et al. 2013. High incidence of antibiotic multi-resistant bacteria in coastal areas dedicated to fish farming. 70(1–2):197–203. doi: 10.1016/j.marpolbul.2013.02.037
Lemery J, Auerbach P. 2017. *Enviromedics: The Impact of Climate Change on Human Health:* Rowman & Littlefield. Lanham, MD.
Levy SB. (1992). Active efflux mechanisms for antimicrobial resistance. *Antimicrobial Agents and Chemotherapy*, *36*(4), 695–703.

Levy SB, Marshall B. 2004. Antibacterial resistance worldwide: Causes, challenges and responses. *Nat Med.* 10:122–129.

Lewbart GA. (2001, January). Bacteria and ornamental fish. In *Seminars in Avian and Exotic Pet Medicine* (Vol. 10, No. 1, pp. 48–56). WB Saunders.

Lin JT, Connelly MB, Amolo C, et al. 2005. Global transcriptional response of *Bacillus subtilis* to treatment with subinhibitory concentrations of antibiotics that inhibit protein synthesis. *Antimicrob Agents Chemother.* 49(5):1915–1926. doi: 10.1128/AAC.49.5.1915-1926.2005

Lisete-Torres P, Losada-Barreiro S, Albuquerque H, et al. 2012. Distribution of hydroxytyrosol and hydroxytyrosol acetate in olive oil emulsions and their antioxidant efficiency. *J. Agric. Food Chem.* 60(29):7318–7325. doi: 10.1021/jf301998s

MacFadden DR, McGough SF, Fisman D, et al. 2018. Antibiotic resistance increases with local temperature. *Nat Clim Chan.* 8:510–514.

Marcos-López M, Gale P, Oidtmann BC, Peeler EJ. 2010. Assessing the impact of climate change on disease emergence in freshwater fish in the United Kingdom. *Transboundary and Emerging Diseases, 57*(5), 293–304.

Markham KR, Andersen OM. 2006. *Flavonoids: Chemistry, Biochemistry, and Applications.* CRC Press, Boca Raton, FL.

Marti E, Variatza E, Balcazar JL. 2014. The role of aquatic ecosystems as reservoirs of antibiotic resistance. *Trends Microbiol.* 22:36–41. doi: 10.1016/j.tim.2013.11.001

Martinez JL. 2014. General principles of antibiotic resistance in bacteria. *Drug Discov Today.* 11:33–39. https://doi.org/10.1016/j.ddtec.2014.02.001

McGough SF, MacFadden DR, Hattab MW, et al. 2020. Rates of increase of antibiotic resistance and ambient temperature in Europe: A cross-national analysis of 28 countries between 2000 and 2016. *Eurosurveillance.* 25(45). doi: 10.2807/1560-7917.es.2020.25.45.1900414.

McPhearson RM, et al. 1991. Antibiotic-resistance in gram-negative bacteria from cultured catfish and aquaculture ponds. *Aquaculture.* 99(3–4):203–211. https://doi.org/10.1016/0044-8486(91)90241-X

Mohanty BR, Sahoo PK. 2007. Edwardsiellosis in fifish: A brief review. 32(7). doi: 10.1007/s12038-007-0143-8

Molbak K. 2005. Human health consequences of antimicrobial drug-resistant Salmonella and other foodborne pathogens. *Clin Infect Dis.* 41(11):1613–1620. doi: 10.1086/497599

Neimann J, Engberg J, Molbak K, et al. 2003. A case-control study of risk factors for sporadic campylobacter infections in Denmark. *Epidemiol Infect.* 130(3):353–366. https://doi.org/10.1017/s0950268803008355

Oldfield E, Lin FY. 2012. Terpene biosynthesis: Modularity rules. *Angew Chem Int Ed Engl.* 51(5):1124–1137. doi: 10.1002/anie.201103110.

Picher Pichersky E, Raguso RA. (2016). Why do plants produce so many terpenoid compounds? *New Phytologist, 220*(3), 692–702. https://doi.org/10.1111/nph.14178

Pachauri RK, Meyer LA. 2014. *Climate Change 2014: Synthesis Report.* Intergovernmental Panel on Climate Change, Geneva, Switzerland. https://www.ipcc.ch/site/assets/uploads/2018/02/SYR_AR5_FINAL_full.pdf

Panche AN, Diwan AD, Chandra SR. 2016. Flavonoids: An overview. *J Nutr Sci.* 5:e47. doi: 10.1017/jns.2016.41

Payne JL, Wagner A. 2019 The causes of evolvability and their evolution. *Nat Rev Genet.* 20:24–38. doi: 10.1038/s41576-018-0069-z.

Pichersky E, Raguso RA. 2018.Why do plants produce so many terpenoid compounds?. *New Phytol.* 220(3):692–702. doi: 10.1111/nph.14178

Ramirez MS, Tolmasky ME. 2010. Aminoglycoside modifying enzymes. *Drug Resist Updat.* 13(6):151–171. doi: 10.1016/j.drup.2010.08.003

Roberts MC. (1996). Tetracycline resistance determinants: mechanisms of action, regulation of expression, genetic mobility, and distribution. *FEMS Microbiology Reviews, 19*(1), 1–24.

Roberts MC. (2005). Tetracycline resistance due to ribosomal protection proteins. *Frontiers in Antimicrobial Resistance: a Tribute to Stuart B. Levy*, 19–28.

Roberts MC. (2012). Mechanisms of bacterial antibiotic resistance and lessons learned from environmental tetracycline-resistant bacteria. *Antimicrobial Resistance in the Environment*, 93–121.

Robinson TP, Pozzi F. 2011. *Mapping Supply and Demand for Animal-Source Foods to 2030*, Animal Production Health Working Paper; Food and Agriculture Organization of the United Nations (FAO). Rome, Italy.

Rodríguez-Verdugo A, Gaut BS, Tenaillon O. 2013. Evolution of *Escherichia coli* rifampicin resistance in an antibiotic-free environment during thermal stress. *BMC Evolutionary Biology.* 13:50. doi: 10.1186/1471-2148-13-50.

Rodríguez-Verdugo A, Tenaillon O, Gaut BS. 2016. First-step mutations during adaptation restore the expression of hundreds of genes. *Mol Biol Evol.* 33:25–39. doi: 10.1093/molbev/msv228

Ruiz J. (2003). Mechanisms of resistance to quinolones: target alterations, decreased accumulation and DNA gyrase protection. *Journal of Antimicrobial Chemotherapy*, *51*(5), 1109–1117.

Salyers AA. (1990). Activities of polysaccharide-degrading bacteria in the human colon. In *Dietary fiber: chemistry, physiology, and health effects* (pp. 187–194). Boston, MA: Springer US.

Schar D, Sommanustweechai A, Laxminarayan R, et al. 2018. Surveillance of antimicrobial consumption in animal production sectors of low- and middle-income countries: Optimizing use and addressing antimicrobial resistance. 15(3). doi: 10.1371/journal.pmed.1002521

Schwarz S, Kehrenberg C, Doublet B, Cloeckaert A. (2004). Molecular basis of bacterial resistance to chloramphenicol and florfenicol. *FEMS Microbiology Reviews*, *28*(5), 519–542.

Shaw WV, Leslie AGW. (1991). Chloramphenicol Acetyltransferase. Annual Review of Biophysics and Biophysical Chemistry, *20*(1), 363–386. DOI: 10.1146/annurev.bb.20.060191.002051

Smith KE, Besser JM, Hedberg CW, et al. 1999. Quinolone-resistant *Campylobacter jejuni* infections in Minnesota, 1992–1998. *N Engl J Med.* 340(20):1525–1532. https://doi.org/10.1056/NEJM199905203402001

Speer BS, Salyers AA. (1989). Novel aerobic tetracycline resistance gene that chemically modifies tetracycline. *Journal of Bacteriology*, *171*(1), 148–153.

Speer BS, Shoemaker NB, Salyers AA. (1992). Bacterial resistance to tetracycline: mechanisms, transfer, and clinical significance. *Clinical Microbiology Reviews*, *5*(4), 387–399.

Springmann M, Godfray HCJ, Rayner M, Scarborough P. 2016. Analysis and valuation of the health and climate change cobenefits of dietary change. 113(15). https://doi.org/10.1073/pnas.1523119113

Stalder T, Barraud O, Casellas M, et al. 2012. Integron involvement in environmental spread of antibiotic resistance. *Front Microbiol.* 3(119). https://doi.org/10.3389/fmicb.2012.00119

Suweis S, Carr JA, Maritan A, Rinaldo A, D'Odorico. 2015. Resilience and reactivity of global food security. 112(22). https://doi.org/10.1073/pnas.1507366112

Taylor DE, Chau A. (1996). Tetracycline resistance mediated by ribosomal protection. *Antimicrobial Agents and Chemotherapy*, *40*(1), 1–5.

Taylor NG, Verner-Jeffreys DW, Baker-Austin C. 2011. Aquatic systems: Maintaining, mixing and mobilising antimicrobial resistance? *Trends Ecol Evol.* 26:278–284. doi: 10.1016/j.tree.2011.03.004

Thilsted SH, et al. 2016. Sustaining healthy diets: The role of capture fisheries and aquaculture for improving nutrition in the post-2015 era. 16:126–131. https://doi.org/10.1016/j.foodpol.2016.02.005

Tuli HS. 2019. *Current Aspects of Flavonoids: Their Role in Cancer Treatment.* Springer, Singapore. https://link.springer.com/book/10.1007/978-981-13-5874-6

UN. The Sustainable Development Goals Report. (United Nations, New York, 2018). https://unstats.un.org/sdgs/files/report/2018/thesustainabledevelopmentgoalsreport2018- en.pdf

Van Boeckel TP, et al. 2015. Global trends in antimicrobial use in food animals. 112(18). https://doi.org/10.1073/pnas.1503141112

Vezzulli L, Grande C, Reid PC, et al. 2015. Climate influence on Vibrio and associated human diseases during the past half-century in the coastal North Atlantic. Proceedings of the National Academy of Sciences, *113*(34). https://doi.org/10.1073/pnas.1609157113

World Bank. 2013. Fish to 2030: *Prospects for Fisheries and Aquaculture (English).* Agriculture and Environmental Services Discussion Paper; No. 3, Washington, DC.

Wright GD. 2007. The antibiotic resistome: The nexus of chemical and genetic diversity. *Nat Rev Microbiol.* 5(3):175–186. doi:10.1038/nrmicro1614

3 Sustainable Antimicrobial Finishes for Textiles

S. Grace Annapoorani and K. Saranya

3.1 INTRODUCTION

Microorganisms like bacteria, yeast, fungi, moulds, and mildew are a part of our daily existence. Microorganisms can be of both healthy and harmful varieties. Microorganisms, which number in the thousands, are present everywhere in the environment and on our bodies. All types of product producers, retailers, and consumers are impacted by these organisms. The breadth of this extends to entire structures, building components, personnel, machinery, processes, textile production, storage, and transportation, as well as textile consumers (Maria C. Thiry, 2010). If one has a thorough understanding of what microbes are, where they come from, and why they thrive on particular materials, one may control their harmful impacts. With the correct technology, this control capacity can offer a useful function on a variety of materials. As long as there is a desire for high-performance fabrics that are pleasant to the skin and have a fresh scent, antimicrobial textile goods will become more and more common. Sports textile is one of several specialized applications that call for contemporary performance materials. These must perform at a high level in terms of longevity, durability, and providing the cloth antibacterial qualities. These qualities can be enhanced, making them more comfortable to wear while also raising the level of hygiene. By reducing skin issues brought on by microbial development and neutralizing odour, treated products are made to be more hygienic. Fabrics typically contain a variety of germs that can harm the wearer in a number of ways (Sharadha Shinde, 2010). The textile industry's rapid expansion has opened up numerous possibilities for a range of cutting-edge finishes, including antimicrobial ones. Natural, renewable resources are becoming more and more important in today's global community as a result of people's commitment to environmental preservation. Value-added finishing increases the fabric value in addition to new developments made utilizing the fabric. The cloth gains advantageous properties like resistance to fire, wrinkles, mildew, etc., during the finishing process. A demanding consumer market has been created by these high value-added fabrics. People are now more conscious than ever before of how important it is to live in a healthy and safe environment. Diseases typically spread from person to person and through surface contact with an infected person's hands, clothes, etc.

The use of antimicrobial materials in baby clothing, underwear, and surgical attire is widespread. Traditional apparel and household textiles are now included in the antimicrobial finishing process. To limit the impact of diseases, antimicrobial agents either kill them or prevent their growth. Because they include carbohydrates, natural fibres like cotton are more quickly damaged by bacteria (R. Rathinamoorthy et al. 2001 ; Krishnaveni V et al. 2007). Numerous industries use antimicrobial-finished fabrics, including sportswear, footwear, furniture, automotive textiles, intimate apparel, and the textiles used in healthcare. A microorganism's presence in the fabric results in an offensive odour, discolouration, and health issues. The clothing that is worn adjacent to the skin must have an antimicrobial finish since microbial illnesses can be harmful to the skin. The application of an antibacterial finish using some herbs is done to protect the wearer's skin.

DOI: 10.1201/9781003268925-3

Numerous products, such as underwear, baby clothing, and surgical gowns, frequently include antimicrobial fabrics. Traditional apparel and household textiles are now being treated with antimicrobial substances. Antimicrobial substances destroy bacteria or stop them from growing, and their effects are controlled. Cotton and other natural fibres are easily destroyed by bacteria because the fibres contain carbohydrates. Athletic equipment, footwear, furniture, medical textiles, automotive textiles, intimate apparel, and other products all use fabrics with antimicrobial finishes (Sathianarayanan M.P. et al. 2010).

3.2 MICROBES OR MICROORGANISMS

Microbes are some of the smallest living things and cannot be seen with the naked eye. Numerous microorganisms, such as bacteria, fungi, algae, and viruses, are carried by microbes. Bacteria are single-celled organisms that expand quickly in warm, damp environments. The bacteria family is further divided into spore-bearing and non-spore-bearing types, as well as Gram-positive (*Staphylococcus aureus*) and Gram-negative (*Escherichia coli*) subgroups. Certain kinds of bacteria are harmful and can spread infection to other people. Moulds, fungi, and mildew are intricate organisms that grow slowly. They damage the fabric's performance characteristics and discolour it. At a pH of 6.5, fungi become active (D. Gupta, et al. 2007).

Commonly found bacteria, fungi, and algae are microorganisms. To thrive and leave heavier stains on the materials, algae need a constant supply of water and sunlight. Between a pH of 7.0 and 8.0, algae are active. Eight-legged dust mites live in home fabrics like blankets, bed linen, pillows, mattresses, and carpets (Rakesh Goyal and C. N. Prabhu, 2009). The freed waste products from the dust mites, which feed on human skin cells, might result in allergic reactions and respiratory problems.

3.2.1 Sources of Microbes

- In breathing air
- In land
- In our skin and bodies

3.2.2 Microbial Growth Ideal Conditions

- In the food we eat
- In warm environments
- In humid surroundings
- On sensitive surfaces (skin, fabric)

3.3 TYPES OF BACTERIA

Numerous creatures, including viruses, bacteria, single-celled plants and animals, certain algae, and fungi, are considered microorganisms. There are harmful microorganisms that can infect humans as well. Scientist Hans Christian Gram divided bacteria into Gram-positive and Gram-negative types based on the structural variations in their cell walls (Sunny Pannu, 2013). Chemical reactions occur as a result of the enzymes found in the cell wall. *Staphylococcus aureus* (*S. aureus*) and *Bacillus subtilis* (*B. subtilis*) are two examples of Gram-positive bacteria that are frequently responsible for food poisoning, sinusitis, and skin diseases like abscesses (Ammayappan, L. and Jeyakodi Moses, J., 2009). *Escherichia coli* and *Pseudomonas aeruginosa* are examples of Gram-negative bacteria (*E. coli*) shown in Figure 3.1. Table 3.1 outlines the distinctions between Gram-positive and Gram-negative bacteria.

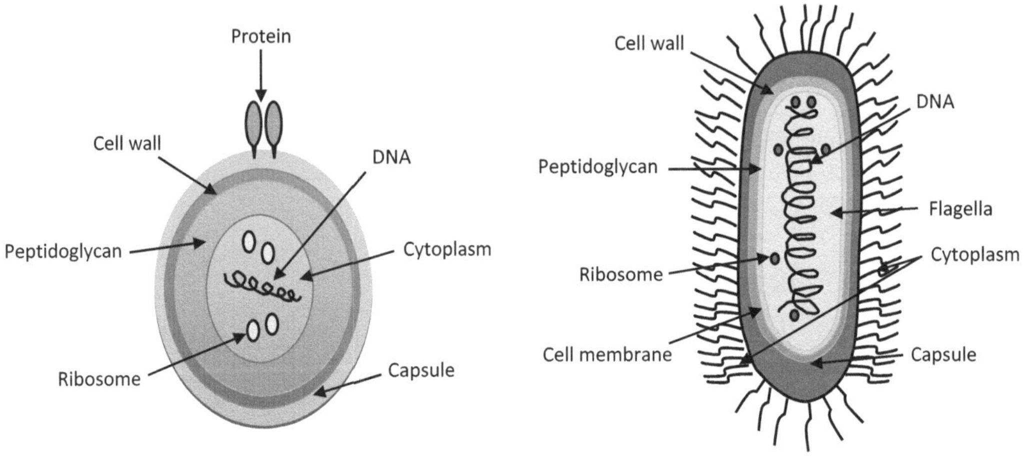

FIGURE 3.1 (*S. aureus*) Gram-positive bacteria and (*E. coli*) Gram-negative bacteria.

3.4 GREEN FINISHING CHALLENGES

In general, natural extracts like chitosan that were tested for antimicrobial activity in textile finishing revealed a number of issues, including extraction issues, the isolation of bioactive compounds, and the strong astringent and bitter smell of some natural chemicals. The treated textile's physical characteristics and other performance characteristics must be preserved during antibacterial finishing (Veeresh Mudnoor and S. K. Laga, 2012). The stiffness and drape of the fabric are significantly influenced by several factors, such as bending rigidity and bending modulus. For instance, after coating the surface with chitosan, the textile's air permeability is decreased, which has an impact on the wearer's comfort. The main issue with fabrics treated with bioactive agents is the poor finish durability during usage and washing. Furthermore, their treatments for textiles work well at the highest concentrations. Nevertheless, these extract formulations are appealing because they are non-toxic and environmentally beneficial (Sanjay R. Malpani, 2013). Further study in the area of bioactive textiles made from natural fibres is needed to provide a workable alternative to antimicrobial textiles based on synthetic material in order to address all of these issues.

3.5 EVALUATION OF ANTIMICROBIALS (ANTIBACTERIAL TESTS)

Several test methods have been developed to gauge the efficiency of antimicrobial fabrics.

TABLE 3.1
Differences between Gram-Positive and Gram-Negative Bacteria

Gram-Positive Bacteria	Gram-Negative Bacteria
A cytoplasmic membrane, amino bridges connecting polymeric layers of peptidoglycan, and a flexible outer layer known as the capsule make up cell walls.	Possess a plasma membrane bilayer, an exterior membrane bilayer, and a thin coating of peptidoglycan.
Due to the dense peptidoglycan multilayer and in accordance with a thin peptidoglycan layer, it maintains violet crystals.	It doesn't maintain the violet dye and becomes pink in colour.
Consist of proteins, teichoic acids, and peptidoglycan. A chemical analysis of the cell wall reveals that peptidoglycan makes up more than 70% of its weight and is covalently linked to teichoic acids by way of a phosphodiester linkage.	The DNA on one chromosome and a plasmid are all that a cell needs to maintain a complex metabolism that allows it to grow and divide.

3.5.1 Quantitative Antimicrobial Tests

These tests allow for the counting of the number of microorganisms on the completed fabrics, and the results are displayed as a percentage or as a log reduction. The AATCC TM100, JIS L1902, AATCC90 percentage reduction, and ISO 20743 shaking flask reduction procedures are test methods for quantitative determination. The majority of tests can evaluate both Gram-positive and Gram-negative bacteria, even though the kinds of microbiological testing that can be done using a particular standard approach vary (K. Selva Subha et al., 2007).

3.5.2 AATCC 100 (Suspension Test)

It is a quantitative antimicrobial standard approach for evaluating the effectiveness of textiles and fabrics against common bacteria over a 24-hour contact period, including *S. aureus*, *Klebsiella pneumonia*, and *E. coli*. It assesses the bacteriostatic (inhibition of bacterial proliferation) and bactericidal (bacteria-killing) activities of the antimicrobial test agent. To get the highest concentration of test bacteria, a popular procedure is to incubate a bacterial sample in growth enrichment broth for 24 hours. High concentrations of germs are introduced into samples of antimicrobial test fabric and untreated control fabric (Wayne Swofford, 2010).

Traditional microbiological testing methods count the quantity of bacteria on the fabrics at the beginning of the contact time (from 0 hours). Following that, the textiles with organisms inoculated on them are incubated for 24 hours at the right temperature and nutrient levels. A lack of antibacterial fabric allows bacteria to grow and multiply. The remaining bacterial counts are noted after neutralization and extraction, and a percent reduction is calculated using the starting count and subsequent count data (W. D. Schindler and P. J. Hauser, 2004). The following equation gives the percentage of antimicrobial elimination:

$$\%R = \frac{(A-B)}{A} \times 100$$

where A is the number of bacteria colonies and B is the number of bacteria on the treated fabric samples.

3.5.3 Qualitative Antimicrobial Tests

These evaluations are qualitative, and the outcomes are expressed as estimated inhibition zones or selected scores. The crucial assessment methods JIS L 1902 and ISO 20743 contain components for qualitative evaluation of antibacterial activity in finished textiles. Similar to these methods, the AATCC TM 147 utilizes the agar diffusion test.

3.5.4 AATCC 147 (Agar Diffusion Test)

The antibacterial activity of treated fabrics against Gram-positive and Gram-negative bacteria is evaluated using a qualitative antimicrobial technique. Treated and untreated samples of the same fabric are positioned in close proximity to growing agar that has already been contaminated with the test organism. The plates are checked for bacterial growth beneath the textiles and around their edges after 18 to 24 hours of 37°C incubation. The size of the inhibition area that forms when the antimicrobial substance diffuses into the agar or the rate at which the active ingredient is produced depends on how well the antimicrobial action works (AATCC Test Method 100-2004, 2012). The main issues with these evaluations (qualitative tests) are their lack of reproducibility and, frequently, a discrepancy between experimental results and the actual state of the field. Time delay, tedious design, high costs, and the test evaluations' reliance on the operator's explanations are some of the drawbacks associated with the use of quantitative methods of assessment. These tests require meticulous attention to detail and well-trained laboratory staff to provide reliable and repeatable results (AATCC Test Method 30-2004. 2012; AATCC Test Method 90-2011, 2012).

3.6 ANTIMICROBIAL TESTING IN TEXTILES

Testing is a helpful tool for textile distribution, consumption, and manufacture. Testing for such chemical compounds that affect an individual's health (particularly skin) is essential because textiles come into close contact with our bodies. This is particularly true for infants and children, because their skin is far more delicate than that of grownups. Testing of textiles can be used to evaluate the effectiveness of antibacterial treatments, the durability of final products, the level of bacteria resistance, and other factors. Assessment procedures carried out under carefully monitored conditions with repeatable outcomes are required to gauge the effectiveness of textiles treated with antimicrobials (Naresh M. et al. 2011; Wayne Swofford, 2010). Some of the available standards are shown in Figure 3.2.

3.7 ANTIMICROBIAL FINISHES AND ITS NECESSITY

To achieve the following goals, textile materials must be treated with an antimicrobial agent.

- To keep microbes under control.
- To safeguard against viral infections brought on by bacterial infections.
- To speed up microbe degradation to cut down on odour production.
- To maintain the quality of textile materials from unforeseen stains, colour fading, and quality degradation (Scalbert, A. 1991).

3.8 REQUIREMENTS FOR ANTIMICROBIAL FINISH

As a result of the chemical processing and the duration of their use, clothes are expected to experience additional wear and tear. For the finish to work as intended, the following requirements must be met.

- It shouldn't be harmful to the supplier, the end user, or the ecosystem.
- It should be durable enough to withstand washing, dry cleaning, and hot pressing.
- Compatibility with different chemical methods.
- Application simplicity.
- Sufficient fabric durability retention.
- To keep bodily fluids away.
- To avoid sterilization and disinfection.

3.9 ANTIMICROBIAL TECHNOLOGY BASED ON PURPOSE OF APPLICATION

Choosing the appropriate antimicrobial technology for the nonwovens, wovens, or composite fabric business for antimicrobial applications is one of the numerous difficulties encountered. Considerations for application should take into account:

- **Durability:** Fabrics that last need components that last. Industrial fabrics with medical facility applications need to have antimicrobial finishes that can withstand abrasion, sterilization, wet/dry cycles, freeze/thaw cycles, alcohol rinse, and other physical and chemical challenges.

FIGURE 3.2 Antimicrobial finishing of textiles with conventional standards.

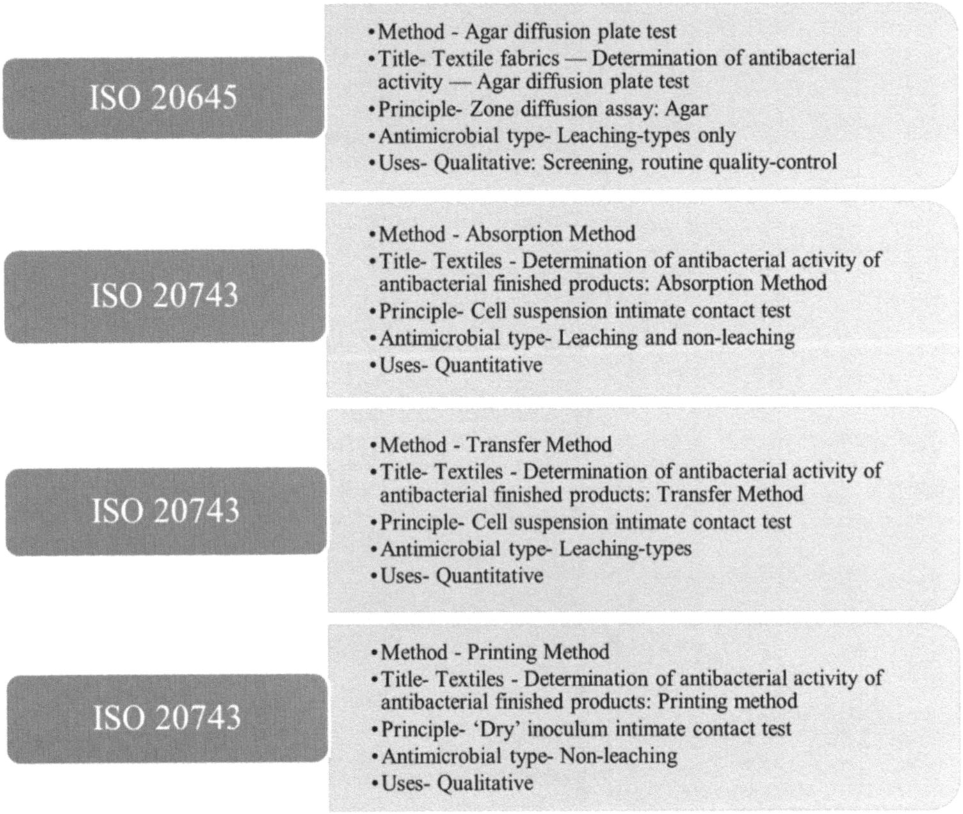

FIGURE 3.2 Continued

- **Waste control/toxicity:** Antimicrobials are effective at eradicating a variety of microbial pests, but they must be used with care to avoid harming beneficial organisms.
- **Spectrum of activity:** Many materials are antibacterial at the proper concentrations, but it is crucial for healthcare applications to have the broadest spectrum of action possible while still remaining safe and effective. This is especially crucial when antimicrobial treatments are incorporated into durable items.
- **Adaptation:** Any soluble substance that has an impact on a microorganism's existence has the ability to create situations in which the microorganism's cellular units adapt or change into resistant forms. When using common sanitizers or disinfectants, a rinse is required after the desired contact time. This is done to reduce the dangers of sub-lethal doses of the antibiotic being present and running the risk of adaptation or other types of resistance (Russell A.D. et al. 1997).

3.10 ANTIMICROBIAL AGENT

- Antimicrobial agents, often known as antimicrobial medicines, are substances that have the ability to either kill or halt the growth of germs. Antimicrobial medications are grouped based on the principal pathogens they target, such as bacteria and viruses. Based on the many chemical components they contain, antimicrobial agents are categorized into two classes. The first category consists of chemical antimicrobial agents, such as antibiotics, as well as metal and metal oxide nanoparticles, such as silver and silver oxide. Herbal antibacterial agents make up the second category.

Chemical antimicrobial

- Antibiotics
- Metal oxide Nano particles
- Metal Nano particles

Herbal antimicrobial

- Lemon balm, eucalyptus, primrose
- Ginger, Mint, tribulus
- Garlic, Portulaca, Clove, Thyme
- Turmeric, Pennyroyal, Chamomile, Cinnamon, Mallows

FIGURE 3.3 Classification of antimicrobial agents according to their various ingredients.

- While playing a significant antibacterial role, antibiotics and other chemical antimicrobial agents can have a number of negative side effects. The production of free oxygen radicals (ROS) is one of the primary adverse effects. ROS are extremely hazardous, and they are believed to be a major factor in the development of cancer.
- Clove, portulaca, tribulus, eryngium, cinnamon, turmeric, ginger, thyme, pennyroyal, mint, fennel, chamomile, burdock, eucalyptus, primrose, lemon balm, mallows, and garlic are among the herbal antibacterial agents that belong to the second category. These biomaterials have the capacity to neutralize free radicals, hence preventing the generation of ROS (Hammer, K. A. et al. 1999) (Figure 3.3).

3.10.1 Properties of an Effective Antimicrobial Finish

The following characteristics of an effective antimicrobial finish should be present:

- Because the rate of growth of germs is very quick, antimicrobial finishes must act immediately to be efficient.
- The antimicrobial substance must prevent the development of germs or destroy them entirely, and it must retain this power even after frequent cleanings or exposure to the environment.
- The antimicrobial must really be secure for both application by the supplier and wear by the customer.
- The finish must adhere to tight government rules and have no influence on the environment.
- The antimicrobial finish should not negatively affect the other fabric attributes, such as wear comfort, and it should be simple to apply and compatible with other finishing agents.
- The antibacterial finish must be reasonably priced.

3.11 EFFECTS OF HERBAL TEXTILE IN SUSTAINABLE WAY

Garden plants known as herbs are raised and harvested for use in food, fragrance, medicine, and fibres. For their distinctive scents, appealing textures, appealing colours, and range of domestic purposes, plant herbs are planted in gardens. Inherent in eco-friendly natural dyes have seen a worldwide renaissance as a result of current eco-consciousness. The use of herbal goods has changed the way that textile items are used to treat a variety of ailments. Additionally, the herbs do not contaminate nearby water sources, which would otherwise pollute the environment. With the use of the herbs, one can produce a wide range of reds, yellows, browns, oranges, and greens.

Natural dyes have been popular for functional finishes that support an eco-friendly way of living. Natural dyes are renewable materials, non-toxic, non-carcinogenic, and environmentally beneficial.

FIGURE 3.4 Herbs used for various disorders.

Plant species of more than 500 that produce natural colours are gifts to the biosphere. The dyes are produced using 200 natural plants, flowers, roots, and barks, as well as their leaves, seeds, and barks. In herbal textiles, no extra colourants are employed; the colour is obtained solely from the medicinal mixture. Its characteristics will therefore endure for as long as the hue does. Herbal clothing is also becoming increasingly fashionable because of its aristocratic simplicity and gorgeous hues created by the herbs.

When dyeing herbal fabrics, only chemical-free herbal preparations are utilized; however, when dyeing with vegetable dyes, chemicals like copper sulphate and ferrous sulphate are employed as catalysts. Because they also impart some therapeutic benefits, herbal dyes differ from vegetable colours in this way. The immune system is supposedly strengthened, and the body systems are restored to equilibrium by wearing herbal clothing. The dyed cloth or yarn has a mystical medicinal effect related to some of the medi-herbs utilized in herbal textiles. The disorders treated by various herbs are listed in Figure 3.4.

3.12 SUSTAINABLE TEXTILES FINISHED WITH SOME OF MEDICINAL PLANTS

Many plant parts, including the rhizome, leaves, fruits, and flowers, are used to extract colour materials. Over 2,000 pigments are thought to be present in different plant sections, but only about 150 have been used economically. The process of making herbal clothing involves using specific herb extracts to colour cotton, silk, linen, and other fibres, as well as yarn and fabric. At no point are synthetic or artificial colours utilized, and particular products are used for washing.

Herbal dyeing was developed through careful study of the traditional dyeing techniques used since the time of the Indus culture. The raw materials for making herbal textiles are 100% handwoven, eco-friendly cotton, silk, wool, linen, jute, hemp, and other natural blends, which are then specially coloured with herbal infusions for both aesthetic and medicinal reasons. While dyeing and processing the fabric, no chemical procedures are used. Before becoming completed goods, the fabrics must go through a number of processes. Herbal fabrics are especially smooth and conducive to perspiration, which aids in disease recovery.

Each cloth has a unique blend of herbs that are known to be disease-curing. The fabric may be used to treat various kinds of illnesses, including eczema and psoriasis. It is also used to treat blood pressure, diabetes, rheumatism, arthritis, heart conditions, and respiratory problems like asthma and sleeping problems. Additionally, it improves immunity. Herbal-medicated clothes can also be used as dressings and bandages due to their antibacterial and anti-inflammatory characteristics. Moreover, it can strengthen the immune system, facilitate weight loss, and ease common body issues. It is also used as a blood purifier, a relaxing agent, an energy booster, a mood enhancer, and a cooling agent. The optimum time to wear clothing that has been herbally treated is while you are resting, which can happen throughout sleep because this is when the body heals and restores balance. Thus, bed linens, towels, clothing for meditation, and cotton mats make up the majority of herbal clothing products (Gupta. D. et al. 2006).

3.13 SUSTAINABILITY IN HERBAL DYEING PROCESS

Beginning with pure organic yarn or fabric, the manufacture of herbal textiles is painstakingly monitored at every stage without the use of any chemical dyes or processing methods. The procedure begins with several phases of treatment on the grey fabric before it is coloured and made fit for use. The use of herbal dyes dates back to the Indus culture and was created after significant investigation. Herbal clothing preparation does not include the use of machines; instead, large vessels made to the required specifications are used for all procedures, including bleaching, dyeing, and other processes. Additionally, no chemicals are used in the spinning, weaving, or aesthetic enhancement processes.

The most widely used fabric is made of organic cotton, which is grown without the use of chemical fertilizers and pesticides. Herbal clothing can also be made from silk, wool, coir, linen, hemp, banana, nettle, and bamboo. Desizing is the first step in the handloom process, which begins with cotton yarn (to remove loose particles, debris, etc.) (Deepti Gupta and Somes Bhaumik, 2006; Pushpalatha K., et al., 2013). The next paragraphs go into detail about the typical procedures used in the production of herbal clothes.

3.13.1 Desizing

The processed grey fabric is washed in mineral-rich water and sea salt to remove the sizing, gums, and oils from the weaving process.

3.13.2 Sustainability in Bleaching

Sun drying and biodegradable organic cleaning agents and surfactants are used to carry out this process. By exposing the fabric to sunshine on a natural grass foundation and utilizing animal manure, you can organically bleach cloth. The cloth or yarn is first bleached with a combination of saptala (*Acacia sinuata*) and phenila, as well as biodegradable, organically generated, organic cleaning chemicals (*Sapindus mukorossi*) (Ganesan P and Vrdhini K.J. 2015). Many Ayurvedic physicians view cow's urine as being extremely good for balancing one's "doshas", or fundamental components of one's physiology and psychology, for boosting one's immune system, and as an elixir for bringing life. The process is completed by directly shining the fabric or yarn in the sun (Chen C. and Chang. WY. 2007).

3.13.3 Sustainability in Mordanting

Natural mordants like myrobalans, rubhab leaves, oils, alum, fruit extracts of haritaki, bark of lodhra, minerals, and iron are utilized to give the colours a vibrant appearance and maintain colour fastness. Because they are not environmentally friendly, heavy metallic mordants like copper, chromium, zinc, and tin should not be used. Before dyeing the fabric or yarn, mordanting is applied to give the colours more vibrancy and keep their fastness (Benjamin Tawaiah et al. 2015).

3.13.4 Dyeing/Medication

Because medical herbs are employed to give fabric or yarn the necessary colour, the phrase medicine is used instead of dyeing. Depending on the condition being treated, the yarn or fabric is next medicated with a carefully curated blend of herbal medicinal remedies. The appropriate herbs are applied directly to the fabric with the aid of natural components so that the therapeutic benefits of the herbs can be imparted. The following are the steps involved:

- The organic cotton yarn and fabric are meticulously coloured using a mixture of herbal dyes that are suggested for treating the targeted condition.
- The therapeutic herbs obtained from plant parts are carefully blended to create the herbal dyes.
- The equipment, blend of herbs, temperature, time, number of soaks, and length are all carefully regulated.
- In order to retain the therapeutic potential of the herbs, desired plants are applied directly to fabric with natural extracts.
- With careful application of herbs, hues of green, red, yellow, brown, and orange can be produced.

- Finally, to remove the non-absorbed particles, the medicated cloth is chilled, rinsed multiple times, and always dried in the shade (Deepti Gupta and Bhaumik S. 2007).

3.13.5 Sustainability in Finishing

After using herbal dyes, the fabric is finished using an all-natural method that includes aloe vera, castor oil, and other natural ingredients, as well as pure water sprinkled on the fabric and pressure stretching.

3.13.6 Recycling Residue

After using herbal dyes, the fabric is finished using an all-natural method that includes aloe vera, castor oil, and other natural ingredients, as well as pure water sprinkled on the fabric and pressure stretching (Gokarneshan N., et al. 2017).

3.14 METHODS FOR HERB TREATMENT IN SUSTAINABLE TEXTILES

Herbal textiles are created by incorporating herbal extract finishes and essential oils into textiles. These textiles have uses in enhancing the scent and therapeutic value of clothing. The incorporation of herbal medicines and oils into textiles involves the use of microencapsulation, sensory perception technologies, liposomes, dyes, and coated textile technology. The finishing temperatures, the quantity and time of plant extract, the combination of herbs, and even the tools used are all carefully regulated.

Some frequently used methods for textile finishing by microencapsulation include spray drying, air suspension coating, solvent evaporation, sol-gel microencapsulation, and *in situ* polymerization. Spray drying is used to encapsulate taste, fragrance, and oils. Core particles are distributed in a polymer solution and sprayed into a heated chamber, where they solidify on the core material to create polynuclear or matrix-type microcapsules. By suspending solid core material particles in a heated or cold, upward-moving stream of air, coating is accomplished (Rajput A., et al. 2017). The coating material can be manufactured from cellulose derivatives, dextrins, thickeners, lubricants, protein derivatives, and starch derivatives. It is atomized through nozzles into the chamber and forms a thin film on the surface of suspended particles. It has been used to give textiles an antifungal finish. A method known as solvent evaporation involves dissolving, dispersing, or emulsifying the drug in an external aqueous or oil phase. Sol-gel microencapsulation is used to create fabrics that are antibacterial, flame retardant, water/oil resistant, UV protective, anti-wrinkle, and self-cleaning. In the process of making a sol-gel, silicon monomer is hydrolyzed, and silica is condensed to create a porous structure with a three-dimensional networked structure that may be tuned for a variety of useful qualities (Ben Fadhel, et al. 2012; Banupriya, J. and Maheshwari, V. 2013).

The process of microencapsulation, which is frequently used in textile finishing, creates miniature capsules with the appropriate qualities utilizing tiny particles or droplets encased in a coating material. The core material of the microcapsule is referred to as such, while the shell refers to the exterior coating. Microcapsules can range in size from a few micrometres to a few millimetres in diameter. By microencapsulating the core substance, several unique and useful features can be added to the fabrics. Any chemical that will perform a specific function for the fabric can serve as this core material. Moisturizers, healing oils, and pesticides can now be incorporated into textiles thanks to encapsulation. Sportswear and medical textiles are also becoming more and more common places to microencapsulate antimicrobial compounds.

The "core material" and the "wall material" should provide a robust finish for the creation of herbal textiles via microencapsulation. The required herbal extract serves as the core ingredient, and this wall could be made of materials like gum acacia. Herbal extracts that have been microencapsulated have demonstrated antibacterial activity and wash durability up to 20 washes (Joshi, M.

et al. 2007). A very popular method for putting fragrance in fabrics is microencapsulation. Under normal circumstances, these attached capsules are unbreakable, and applying pressure on the cloth releases scent, which promotes healing in aromatherapy.

3.15 MICROENCAPSULES FOR ANTIMICROBIAL FINISHES

Bacteria frequently induce fabric degradation, which results in the loss of a variety of valuable characteristics. Antimicrobial coatings, which can be applied with the aid of microencapsulation, can be used to prevent this issue. For fabrics used in technical and medical applications, this finish is very advantageous.

3.16 PAD–DRY–CURE PROCESS

The pad–dry–cure process was used to apply herbal extracts made from chamomile, sage, and green tea to cotton fabric. Gram-negative, Gram-positive, and yeast-resistant bacteria were all resistant to the treatment material's antimicrobial effects. Even after ten cycles of washing, its antibacterial activity was still present.

3.17 ROLE OF NANOTECHNOLOGY IN TEXTILES

In the course of their useful lifetime, microbes like protozoa, algae, fungi, viruses, and bacteria have a greater propensity to damage textiles, especially those made of cellulose fibres like lyocell, viscose, linen, and cotton. The antimicrobial function has recently evolved into a necessary requirement for all clothing, medical textiles, and household products as people become more aware of health and cleanliness. Numerous metal nanoparticles, including silica, titanium, gold, zinc, copper, and silver, as well as metal oxides including copper oxide, zinc oxide, and titanium dioxide, are currently the subject of extensive research as potential antibacterial agents (Ammayyappan L. and Jeyakodi Moses. J. 2009). Compared to conventional antibacterial agents, nanoparticles with a larger surface area-to-volume ratio contribute a superior antimicrobial property. The use of antimicrobial nanostructured materials in textiles has shown positive synergistic antibacterial properties compared to a single nanomaterial. Cotton fibres treated with silver nanoparticles made from extracts of the Chinese holly plant have cost-effective and environmentally friendly antibacterial capabilities (Joshi M., et al. 2010).

3.18 ANTIMICROBIAL-FINISHED SUSTAINABLE TEXTILES

Sustainable fabrics like cotton and wool, which hold onto oxygen, water, and nutrients, are prone to microbial development and even dust mites. Therefore, antimicrobial treatments should be used on textiles to prevent the growth of bacteria, as well as to shield them from odour, loss of colour, and quality degradation. The bed, bath, and kitchen linens, as well as drapes, rugs, cushions, and mattresses, can all become deteriorated by microorganisms. In addition, a number of microbes flourish on skin, and dust mites are attracted to skin cells that are shed on linens, towels, and clothing. An enormous number of textiles with heavy traffic are kept in hospitals. Specific finishes for hospital use are necessary due to the frequent influx of individuals, especially those with contagious diseases. The risk of infection for patients, caregivers, and hospital staff arises from the intrinsic susceptibility of textile fibres to the development of microorganisms.

In the past, natural dyes were effectively used to colour and treat textiles with antibacterial properties at the same time. *Escherichia coli, Staphylococcus aureus, Candida albicans, and Candida tropicalis* were all successfully repelled by the catechu finish applied to wool. Catechu is a promising antibacterial ingredient for the development of bioactive textile materials and herbal apparel, as evidenced by the dye's observed antimicrobial properties and little cytotoxicity (Saravanan M. and

Bharathi Dhurai, 2017). One of the many natural, non-metallic, and antibacterial finishes available is chitosan, a deacetylated version of chitin that is an essential component in crustacean shells and has been shown to be effective against both Gram-positive and Gram-negative bacteria. To increase the antimicrobial effectiveness of the finishes and mitigate the drawbacks of the treatments, a variety of bioactive chemicals are combined in the production of antimicrobial textiles. Combining finishes lowers the likelihood of strains developing that are resistant to medication (Rahman M.A. et al., 2010).

Using pure cotton woven or knitted fabrics treated with the 16 chosen medicinal herb extracts, such as neem, turmeric, holy basil, and sandalwood, seven different diseases have been evaluated for treatment using these materials, including skin rashes, psoriasis, breathing problems, immune disorders, nausea, muscle aches, and sinus trouble/cold. There are now seven main forms of therapeutic clothing. Clinical studies and antibacterial evaluations of fabrics treated with medicinal herb extracts have shown a correlation between the effectiveness of the treatment and its antibacterial activity (Simonic B. and Tosmic B. 2010; Priya Jaiswal, et al. 2017; Mariam Y., et al. 2021). In every case, it was discovered that the clothing's curative ability was significant and persisted for 10 to 15 washes.

The protective qualities of herbal dyes have only recently been discovered, despite the fact that they have long been known for their dyeing and therapeutic worth. Many of the plants used to make dyes are considered medicinal, and some have lately been found to have strong antibacterial properties. Curcumin from turmeric, lawsone from *Lawsonia inermis*, juglone from walnut, lapachol from taigu, and catechin from *Acacia catechu*, as well as a number of anthraquinones from *Rubia tinctorum*, *Rubia cordifolia*, *Rheum emodi*, *Punica granatum*, and *Quercus infectoria*, have all been shown to have antimicrobial activity. Particularly when utilized in close contact, these natural antibacterial medicines may dramatically lower the risk of illnesses. Additionally, natural bioactive substances found in natural dyes and pigments have been described as important antibacterial finishing agents in green dyeing processes (Dahham S.S. et al. 2010).

3.18 CONCLUSION

The textile sector is constantly looking for solutions to this issue after realizing how the usage of hazardous chemicals throughout manufacturing processes contributes negatively to environmental contamination. One crucial step toward adopting an organic lifestyle and lowering environmental pollution is herbal dyeing. Pesticides, fertilizers, antibiotics, growth hormones, genetically modified organisms, chemicals, irradiation, and sewage sludge are absent from herbal textile products, which prevents numerous illnesses that are frequently associated with wearing conventional clothing.

Future predictions indicate that consumers' demand for novel eco materials will rise. In recent years, progress has been made in areas such as the discovery of novel sources, formulations, extraction, purification, and stability procedures in an effort to generate naturally occurring colourants that are economically viable. Investigations into hygiene and environmental safety are required since they have a great chance of leading to the discovery of natural pigments that are comparably better and more stable and could have more industrial uses. This is true despite enthusiastic studies examining the economic viability of commercial natural dye synthesis and use for sustainable bioresource exploitation.

In the coming years, herbal textiles may overtake conventional textiles as a major industry. The best method to embrace a healthy lifestyle is to combine herbs and textiles in an environmentally friendly way. All across the world, there is growing awareness of herbal clothes. One of the best ways to resurrect and grow the Indian handloom industry's market share in the global textile industry is through the use of herbal treated fabrics. Additionally, herbal clothing is eco-friendly, biodegradable, nontoxic, and non-carcinogenic; these qualities make it an easy choice for a modern, educated consumer searching for a healthy product. Furthermore, the need for bio colourants to fulfil the growing demand will compel business owners to enter this market in an effort to increase earnings and create more jobs.

REFERENCES

AATTC Test Method 30-2004. 2012. *Antifungal activity, assessment on textile materials: mildew and rot resistance of textiles.* AATCC Technical Manual, American Association of Textile Chemists and Colorists, Research Triangle Park, NC, 80–82.

AATCC Test Method 90-2011. 2012. *Antibacterial activity assessment of textile materials: Agar plate method.* AATCC Technical Manual, American Association of Textile Chemists and Colorists, Research Triangle Park, NC, 125.

AATCC Test Method 100-2004. 2012. *Antibacterial finishes on textile materials: Assessment of.* AATCC Technical Manual, American Association of Textile Chemists and Colorists, Research Triangle Park, NC, 148–149

Ammayappan, L. and Jeyakodi Moses, J. 2009. Study of antimicrobial activity of Aloe vera, chitosan, and curcumin on Cotton, *Wool, and Rabbit Hair Journal of Fibers and Polymers*, 10(2): 161–166.

Banupriya, J. and Maheshwari, V. 2013. Comparative study on antibacterial finishes by herbal and conventional methods on the woven fabrics, *Journal of Textile Science & Engineering*, 3: 1–12.

Ben Fadhel, B., Assi, A., ladhari, N., Deghrigue, M., Chemli, R. and Joly, J.P. 2012. Antibacterial effects of two Tunisian Eucalyptus Leaf extracts on wool and cotton fabrics, *Journal of The Textile Institute*, 103: 1197–1204.

Chen, C. and Chang, W.Y. 2007: Antimicrobial activity of cotton fabric pretreated by microwave plasma and dyed with onion skin and onion pulp extractions, *Indian Journal of Fibre& Textiles Research*, 1: 122–125.

Dahham, S.S., Ali, M.N. and Tabassum, H. Khan. 2010. Studies on antibacterial and antifungal activity of pomegranate (Punicagranatum L.), *American Eurasian Journal of Agricultural & Environmental Sciences*, 9: 273–281.

Ganesan, P. and Vrdhini, K.J. 2015. Herbal treated microbial resistant fabrics for health care textiles, *Indian Journal of Natural Product Resource*, 6: 227–230.

Gokarneshan, N, Nagarajan, V. B. and Viswanath, S.R. 2017. Development in antimicrobial textiles – some insights on current research trends, *Journal of Scientific & Technical Research*, 1: 230–233.

Goyal, Rakesh and Prabhu, C.N. 2009. Antimicrobial finishing of textiles, *Colourage*, September, 78.

Gupta, Deepti and Bhaumik, Somes. 2006. Review article – Antimicrobial treatments for textiles, *Indian Journal of Fibre & Textile Research*, 32: 254–263.

Gupta, Deepti and Bhaumik, Somes. 2007. Antimicrobial treatments for textiles, *Indian Journal of Fibre and Textile Research*, 32: 254–263.

Gupta, Deepti, Khare, S.K. and Laha, A. 2006. Antimicrobial properties of natural dyes against gram-negative bacteria, *Coloration Technology*: 120–167.

Hammer, K.A., Carson, C.F. and Riley, T.V. 1999. Antimicrobial activity of essential oils and other plant extracts, *Journal of Applied Microbiology*, 86: 985–990.

Jaiswal, Priya, AgyaPreet, Simran and Goel, Garima. 2017. Antimicrobial activity of herbal treated cotton fabric, *International Journal of Engineering and Technology*, 4: 39–43.

Joshi, M., Purwar, R., Wazed Ali, S. and Rajendran, S. 2010. *Antimicrobial textiles for health and hygiene applications based on eco-friendly natural products, medical and health care textiles*, Woodhead Publishing, 84–92.

Joshi, M., Wazed Ali, S. and Rajendran, S. 2007. Antibacterial finishing of polyester/cotton blend fabrics using neem (*Azadirachta indica*): A natural bioactive agent, *Journal of Applied Polymer Science*, 106: 794–800.

Kamel, Mariam Y. and Hassabo, Ahmed G. 2021. Anti-microbial finishing for natural textile fabrics, *Journal of Textiles, Coloration and Polymer Science*, 18(2): 83–95.

Krishnaveni, V., Rajkumar, G. and Shanmugam, S. 2007. Antimicrobial finish in textiles, *Colourage*, 46.

Mudnoor, Veeresh and Laga, S.K. 2012. Application of antimicrobial finish in medical textiles – An Overview, *Colourage*, 59(10): 40.

Naresh, M. Saraf, Sabale, Dr Ashok G. and Rane, Vaishali. 2011. *Antimicrobial finish: Strengthening the next generation*, International Dyer, 35–36.

Pannu, Sunny. 2013. Investigation of natural variants for antimicrobial finishes in innerwear a review paper for promotion of natural hygiene in innerwear, *International Journal of Engineering Trends and Technology*, 4(5): 2168–2170.

Pushpalatha, K., Dr. Anitha, D. and Dr. Sarada, Devi. 2013. Antimicrobial micro encapsulated cellulosic textiles, *International Journal of Science and Research*, 4: 882–884.

Rahman, M.A., Ahsan, T. and Islam, S. 2010. Antibacterial and antifungal properties of the methanol extract from the stem of *Argyreia argentea*, *Journal of Pharmocol*, 5: 41–44.

Rajput, A., Ramachandran, M., Gotmare, V.D. and Raichurkar, P.P. 2017. Resent bioactive materials for development of eco-friendly diapers: An overview, *Journal of Pharmaceutical Sciences and Research*, 9(10): 1844–1848.

Rathinamoorthy, R., Udayakumar, S. and Thilagavathi, G. 2001. Antibacterial efficacy analysis of *Punica granatum L.* leaf, rind and *Terminalia chebula* fruit extract treated cotton fabric against five most common human pathogenic bacteria, *International Journal of Pharmacy & Life Sciences*, October, 2(10): 1147.

Russell, A.D., Furr, R. and Maillard, J.Y. 1997. Microbial susceptibility and resistance to biocides, *ASM News*, 63: 481–487.

Sanjay, R. Malpani. 2013. Antibacterial treatment on cotton fabric from neem oil, aloe vera & tulsi, *International Journal of Advance Research in Science and Engineering*, 2(7): 37–39.

Saravanan, M. and Dhurai, Bharathi. 2017. Comparison of natural agents to develop antibacterial kitchen textiles, *International Research Journal of Pharmacy*, 8: 93–99.

Sathianarayanan, M.P., Bhat, N.V., Kokate, S.S. and Walunj, V.E. 2010. Antibacterial finish for cotton fabric from herbal products, *Indian Journal of Fiber and Textile Research*, 35: 50–58.

Scalbert, A. 1991. Antimicrobial properties of tannins, *Phytochemistry*, 30: 3875–3883.

Schindler, W.D. and Hauser, P.J. 2004. *Chemical finishing of textiles*, Wood Head Publishing, 165–172.

Selva Subha, K.A., Vijay Anand, A., Anita Hebsiba, G. and Thambidurai, S. 2007. Finishing of cotton fabric with seaweed extract for anti-bacterial activity, *Colourage*, November, 80–81.

Shinde, Sharadha. 2010. Significance of microbiological tests in technical textiles, *Man-made Textiles of India*, July, 38(7): 246–247.

Simoncic, B and Tomsic, B. 2010. Structures of novel antimicrobial agents for textiles – a review, *Textile Research Journal*, 80: 1721–1737.

Tawaiah, Benjamin, Badoe, William and Fu, Shaohai. 2015. Advances in the development of antimicrobial agents for textiles: The quest for natural products: Review, *Fibres & Textiles in Eastern Europe*, 24(117): 136–149.

Thiry, Maria C. 2010. Testing antimicrobial performance, *AATCC Review*, 10(6): 26.

Wayne Swofford. 2010. An overview of antimicrobial testing for textile applications, *AATCC Review*, 10(6): 51.

4 Integrated Approach to the Process Design and Experimental Analysis of Lab-Scale Moving Bed Biofilm Reactor using Eco-Friendly Carrier

R. Santhosh Kumar

4.1 INTRODUCTION: BACKGROUND AND DRIVING FORCES

The scientific monitoring of environmental pollution is important for many reasons. It is necessary here to give some idea of the different methods used to stop pollution. Information about the desired limit is essential for calculating the extent to which pollution control is required. Proper monitoring would also help to focus on the seriousness of the problem of environmental pollution and help generate concern among policy planners. A high standard of living involves a high water demand and, at the same time, causes much greater pollution in this essential element. The resultant interference to the natural cycle can often overwhelm processes of recovery, so that, in addition to products arising from the decomposition of natural substances (e.g. proteins, greases, carbohydrates), there is a build-up of anthropogenous additives, such as pesticides, effluents, and garbage, which contaminate drinking water supplies with their toxic or hormonal effects. These contaminants may also consume such large quantities of oxygen that water resources become polluted. To prevent the threat of possible danger to health, or the very existence of certain species, it is essential to determine the quality of a water source before water is drawn off for consumption. The pulp and paper industry is one of the world's oldest and core industrial sectors. The socio-economic importance of paper has its own value to a country's development as it is directly related to its industrial and economic growth.

4.2 EXPERIMENTAL ANALYSIS OF THE FULL-LAB SCALE MOVING BED BIOFILM REACTOR

A crucial factor in the moving bed biofilm reactor (MBBR) process is the sufficient and frequent contact between the wastewater components and the microorganisms attached to the carrier (biofilm). The necessary mixing of the carrier elements in the reactor (reaction tank) can be achieved using the air supply required for aerobic treatment processes. In anaerobic treatment processes, the carriers in the reaction tank can be moved properly by pumping water or by using immersed and slowly rotating mixers. Problems caused by a decrease in biodegradation efficiency are largely a consequence of the intense thickness of the biofilms, resulting in a reduction of the active carrier

DOI: 10.1201/9781003268925-4

surface area. The MBBR carrier does not have this issue. Its process is very different in that the biofilms, which colonize the pores of the carrier and try to grow out of the pores, are kept constantly thin by shear forces. These shear forces evolve on the outsides of the carrier elements due to their movement in the water and provide the self-cleaning effect of the MBBR carrier. Consequently, the biofilms remain thin as they are protected from the shear forces only inside the pores. Thin biofilms are crucial for an optimal supply of microorganisms with substrates (removable pollutants and nutrients) as well as with the oxygen required for the aerobic biodegradation processes.

This research examined full-lab scale MBBR systems treating industrial wastewater generated by paper manufacturing industries in Salem, Tamilnadu. A subsequent study investigated biofilm succession on conventional carriers (AnoxKaldnes™ K1) from initial seeding and compared its efficiencies with the eco-friendly carrier (ridge gourd fibre). Comparisons of the successional development of bacterial (biofilm) communities were also made between the two carriers. The results provided in-depth knowledge of the development of bacterial biofilm of MBBR from initial attachment to maturation. The present study involves the applicability of ridge gourd fibre as an adsorbent (carrier) for the removal of BOD (Biochemical Oxygen Demand) from industrial wastewater. The eco-friendly carriers were prepared from the ridge gourd fruit; as it matures, the fruit becomes increasingly fibrous, which makes it unfit for consumption. A major motivation of this study was to identify the carrier in econocentric and eco-centric points of waste management, as they reduce environmental degradation. Three reactors (MBBR-A, MBBR-B, MBBR-C) were installed on 15 February 2018 by varying the percentage of the eco-friendly carrier fill fraction to 40%, 50%, and 60% volume of the reactor, respectively, with the hydraulic retention time of 15 days. Figures 4.1 and 4.2 show the carrier used for the lab-scale MBBR process.

4.2.1 Conventional Carrier (AnoxKaldnes™ K1)

Carriers are, of course, a vital ingredient in AnoxKaldnes™ MBBR technology. On their protected surfaces different microorganisms ranging from bacteria to ciliates and rotifers are established in a biofilm. These carriers are kept in motion either by the blast air injection in aerobic systems or by stirrers in anoxic or anaerobic systems. Fromthis motion, impurities in the water are transported to the biofilm and are reduced by it.

4.2.2 Eco-Friendly Carrier (Ridge Gourd Fibre)

Ridge gourd fibre originates from the fruit of a sub-tropical vine that belongs to the cucumber family and is native to central and eastern Asia, including the Indian subcontinent. The plant is quite hardy and easy to cultivate and is even grown indoors in regions with colder climates. Ridge gourd

FIGURE 4.1 Eco-friendly carrier (ridge gourd fibre).

FIGURE 4.2 Conventional carrier (AnoxKaldnes™ K1).

is popular as a vegetable in various regional cuisines in Asia, but the fruit is only edible before it ripens. As it matures, the fruit becomes increasingly fibrous, which makes it unfit for consumption. Luffa is a genus of tropical and sub-tropical vines in the cucumber (Cucurbitaceae) family. In everyday non-technical usage, the luffa, also spelt loofah (referred to as gilki or ram tori in India), usually means the fruit of the two species L. aegyptiaca and L. acutangula. The fruit of these species is cultivated and eaten as a vegetable. The fruit must be harvested at a young stage of development to be edible. The vegetable is popular in India, China, and Vietnam. When the fruit is fully ripened, it is very fibrous. Various dishes of gilki are prepared and eaten in India. The fully developed fruit is the source of the loofah scrubbing sponge, which is used in bathrooms and kitchens all over the world. Luffas are not frost-hardy and require 150 to 200 warm days to mature.

4.2.2.1 Function and Role of Carriers

Although carrier surface area has a considerable impact on MBBR performance, other aspects of carrier design should not be neglected. Below follows an overview of different functions related to carrier design, pinpointing the many aspects to be considered when developing new carrier elements as well as when choosing the right carrier type for a certain process.

4.2.2.2 Mixing and Mass Transfer

To keep the carriers in suspension, aeration and/or mechanical mixing are needed in the MBBR. The required mixing intensity and resulting energy requirement for mixing depend on reactor dimensions and volumetric load, as well as on carrier design. The density of the carrier material will also affect the mixing requirements; carriers with a density close to that of water are easier to mix since they are less prone to float on the reactor surface (McQuarrie & Boltz, 2011). Furthermore, the filling degree of carriers in the reactor may affect the mixing pattern, where a reactor containing a lower carrier filling degree will enable better movement of carriers (Barwal & Chaudhary, 2015).

4.2.2.3 Carrier Clogging

An outer fraction of the biofilm is usually active as a result of limited substrate diffusion. When the biofilm is only partially penetrated by substrates, any additional increase in thickness will not have any beneficial effects on the process but rather negative impacts, such as (i) reduced biofilm area, (ii) potentially anaerobic layers causing biofilm sloughing and/or odours, and (iii) increased weight of the carrier (Boltz & Daigger, 2010).

4.2.2.4 Biofilm Attachment and Growth

Inevitably, carriers that effectively retain thin biofilms in larger voids, as a result of high shear forces, are less inclined to support initial biofilm colonization. For some microbial processes, such as anammox, biofilm growth can be very slow, and pre-seeded carriers are sometimes added to the process during start-up to speed up the initial colonization of the system (Christensson et al., 2013).

4.2.2.5 Oxygen Transfer

Aerobic MBBRs require continuous aeration for the supply of oxygen, as well as for keeping the carriers in suspension. Aeration is often a major part of energy consumption at wastewater treatment plants, and the aeration requirement for oxygen supply is directly related to the oxygen mass transfer from air bubbles to the bulk liquid, i.e. the oxygen transfer efficiency (OTE). The OTE depends on bubble retention time, the size of the air bubble (i.e. the area of the gas-liquid interface), and the viscosity of the liquid phase.

4.3 INTEGRATED APPROACH TO THE PROCESS DESIGN OF LAB-SCALE MBBR PROCESS

There are many ways to assess the MBBR process: from studying the biofilm on one individual carrier to evaluating the overall performance of a full-scale reactor. Comparisons between different systems are, however, not always straightforward, owing to the many different functions of MBBR carriers, as discussed above. Below is the lab-scale MBBR methodology used in this research project.

4.3.1 Lab-Scale Reactor

The main aim of the studies presented in this report was to establish trends and principles of the MBBR process, for which a lab-scale setup was sufficient. The majority of studies were performed in three small lab reactors (Figure 4.3). There are several advantages to studying the MBBR process in a scaled-down lab system: the flexibility is high in terms of loading rates and reactor configuration, and the necessary amount of wastewater feed is relatively low. In addition, no restrictions apply regarding effluent quality. Hence, different feeding schemes can be applied using real wastewater, and the reactors can be controlled and monitored to a high level of detail, operating at stable loading and reactor conditions, which cannot be achieved when studying a pilot plant operating on real wastewater. Figure 4.4 shows the working performance of a lab-scale MBBR using an eco-friendly carrier and a conventional carrier (AnoxKaldnes™ K1).

FIGURE 4.3 Pictorial representation of ridge gourd as eco-friendly carrier.

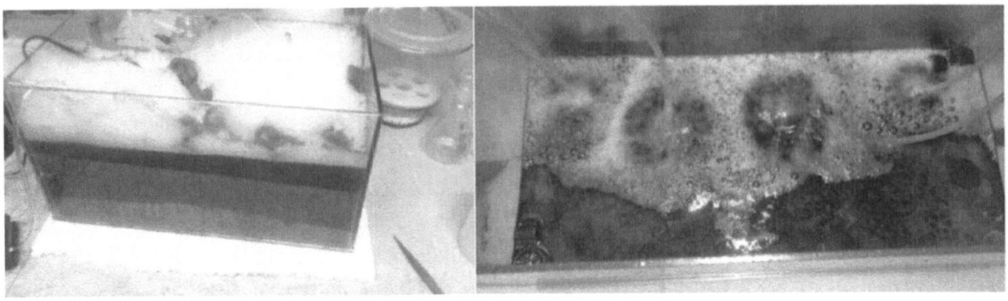

FIGURE 4.4 Working performance of lab-scale MBBR using eco-friendly carrier and conventional carrier (AnoxKaldnes™ K1).

On the other hand, due to the small scale of the systems, possibilities for in-depth microbiological evaluations are limited. Before moving into full-scale applications, it is advisable to test the lab-scale findings in an intermediate pilot-scale setup in order to determine ideal mixing patterns, as well as the system response to varying load and feed composition. The lab-scale reactors using eco-friendly carrier (ridge gourd fibre) have been monitored as the first cycle and, based on the performance of the three reactors, the second cycle was started on 5 March 2018 by varying the percentage of the conventional carrier (AnoxKaldnes™ K1) fill fraction to 40%, 50%, and 60% volume of the reactor.

4.3.2 Comparing Parallel Lab-Scale MBBRs

Many studies presented in this thesis compare the effects of varying the carrier filling degree or the carrier type in parallel MBBRs. In addition, the reactor operation was varied in some of the studies by altering loading rate, feed composition, and HRT and/or dose concentration over time. When varying an operational parameter, such as filling degree, between parallel lab-scale reactors, several factors are affected. Although the reactors may be operated at similar loading rates and mixing intensities, effluent concentrations and reactor hydrodynamics will depend on the activity in the reactors and on the movement of the carriers, which in turn may be affected by carrier design and filling degree. When these conditions differ, substrate availability to the biofilms will vary, resulting in different biofilm performances in the different reactors.

Hence, differences in the overall reactor performance will be related both to differences in carrier area (a direct effect of filling degree) in the systems and to differences in biofilm activity on the carriers (an indirect effect of filling degree). Due to this complex interaction, the interpretation of reactor performance based on a single parameter, such as surface area removal rate, may be ambiguous. Therefore, operational aspects, biofilm characteristics, and reactor conditions have all been considered when comparing results from different reactors in the present work.

4.4 RESULTS AND DISCUSSION

MBBR is used most commonly for contacting wastewater with the carrier. Moving beds can be operated singly, in series, or in parallel. The advantage of flow design is that the adsorption of organics and filtration of suspended solids are accomplished in a single step. Although these moving bed biofilm reactors have been used, the eco-friendly carrier is used more commonly to lessen the chance of accumulating organic material in the carrier. If soluble organic material is not maintained at a high level, more frequent regeneration of the carrier may be required. Lack of consistency in pH, temperature, and flow rate may also affect the performance of contactors.

This research is further accomplished at a stroke of the 7th, 10th, and 15th day of the reactor through biofilm generation. Although moving bed biofilm reactors using conventional carriers have been used, eco-friendly carriers are used to lessen the chance of environmental degradation.

4.4.1 Characteristic of the Carrier

To compare the performance of both carriers under certain operational conditions, each carrier was placed directly in a lab-scale system and used in the present work. The carrier (Figure 5.1) contained separate, individually aerated compartments, which could be filled with different microorganisms to generate the biofilm.

Compared to the operating condition of the conventional carrier (AnoxKaldnes™ K1), the eco-friendly carrier (ridge gourd fibre) was a simple approach to studying growing biofilms on different carrier fill fractions under lab-scale conditions. However, conditions, such as substrate availability and mixing, may differ inside the reactor from those in the lab-scale reactor, as well as between the different carrier compartments. Table 4.1 shows the working performance of a lab-scale MBBR using eco-friendly carrier and conventional carrier (AnoxKaldnes™ K1).

4.4.2 Experimental Result of MBBR Using Eco-Friendly Carrier (Ridge Gourd Fibre)

In the experimental analysis of MBBR using an eco-friendly carrier (first cycle) with an HRT of 15 days, the removal efficiency of the organic pollutant was better when the MBBR was working with a carrier load of 40%, as shown in Table 4.2.

Figure 4.5 shows the biofilm generation of the eco-friendly carrier (ridge gourd fibre) on the 7th, 10th, and 15th days. In addition, the biofilm-shaped carrier only enabled observations under prevailing conditions, and no control strategies could be applied to optimize the performance of the carriers. The activity of biofilm generation in each carrier's compartment could then be visualized and compared at a stroke of the 7th, 10th, and 15th day.

4.4.3 Experimental Result of MBBR Using Conventional Carrier (AnoxKaldnes™ K1)

Table 4.3 describes the experimental analysis of MBBR using a conventional carrier (AnoxKaldnes™ K1). with an HRT of 15 days. The removal efficiency of the organic pollutant was better when the MBBR had been working with a carrier load of 60%.

However, to assess the suitability of both carriers of MBBR processes (e.g. comparing a controlled biofilm thickness to unlimited biofilm growth), the eco-friendly carrier (ridge gourd fibre) is a useful alternative to the conventional Carrier (AnoxKaldnes™ K1).

TABLE 4.1
Characteristic of Eco-Friendly Carrier (Ridge Gourd Fibre) and Conventional Carrier (AnoxKaldnes™K1)

		Eco-Friendly Carrier (Ridge gourd fibre)	Conventional Carrier (AnoxKaldnes™ K1)
Parameters	**Unit**	**Value**	**Value**
Length	Cm	2–3	1.5
Diameter	Cm	3–5	2
Carrier shape	No unit	Circular	Circular
Moisture content	Percentage	58.6	No value
Carrier density	g/cm^3	1	1
Carrier fill fraction	Percentage	40, 50, 60	40, 50, 60

TABLE 4.2
Experimental Result of MBBR Using Eco-Friendly Carrier (Ridge Gourd Fibre) First Cycle

Parameter	MBBR – A (60%)	MBBR – B (50%)	MBBR – C (40%)
pH (no unit)	7.5	8.0	7.0
Temperature (°)	41.0	41.1	41.2
Fluoride (mg/l)	1.5	1.0	1.0
Alkalinity (mg/l)	600	400	350
Chloride (mg/l)	1250	500	4001
Hardness (mg/l)	750	350	300
Total dissolved solids (mg/l)	3120	1500	1260
BOD (mg/l)	862.5	850	659
COD (mg/l)	6979	7018	7380
Total suspended solids (mg/l)	500	423	378
Total volatile solids (mg/l)	3800	3500	3478
Total settable solids (mg/l)	240	178	150

4.5 COMPARATIVE ANALYSIS OF MBBR USING ECO-FRIENDLY CARRIER (RIDGE GOURD FIBRE) – FIRST CYCLE AND CONVENTIONAL CARRIER (ANOXKALDNES™ K1) – SECOND CYCLE

This research investigates the comparison between the performances of a lab-scale Moving Bed Biofilm Reactor with an inflow rate of 8 litres during the hydraulic retention time of 15 days by varying the carrier fill fraction to 40%, 50%, and 60% for both the eco-friendly carrier (ridge gourd fibre) – first cycle and conventional carrier (AnoxKaldnes™ K1) – second cycle.

Case I: Comparative analysis of MBBR – A (60%) performance using eco-friendly carrier (ridge gourd fibre) vs. conventional carrier (AnoxKaldnes™ K1) is shown in Figure 4.6.

Case II: Comparative analysis of MBBR – B (50%) performance using an eco-friendly carrier (ridge gourd fibre) vs. a conventional carrier (AnoxKaldnes™ K1) is shown in Figure 4.7.

TABLE 4.3
Experimental Result of MBBR Using Conventional Carrier (AnoxKaldnes™ K1) – Second Cycle

Parameter	MBBR – A (60%)	MBBR – B (50%)	MBBR – C (40%)
pH (no unit)	8.0	8.0	7.5
Temperature (°)	40.1	41.2	41.9
Fluoride (mg/l)	2.5	2.0	1.5
Alkalinity (mg/l)	1200	700	450
Chloride (mg/l)	1000	950	500
Hardness (mg/l)	900	450	350
Total dissolved solids (mg/l)	2672	2520	1560
BOD (mg/l)	1100	2565	2667
COD (mg/l)	7221	6373	6487
Total suspended solids (mg/l)	414	400	365
Total volatile solids (mg/l)	3600	3569	3029
Total settable solids (mg/l)	209	195	185

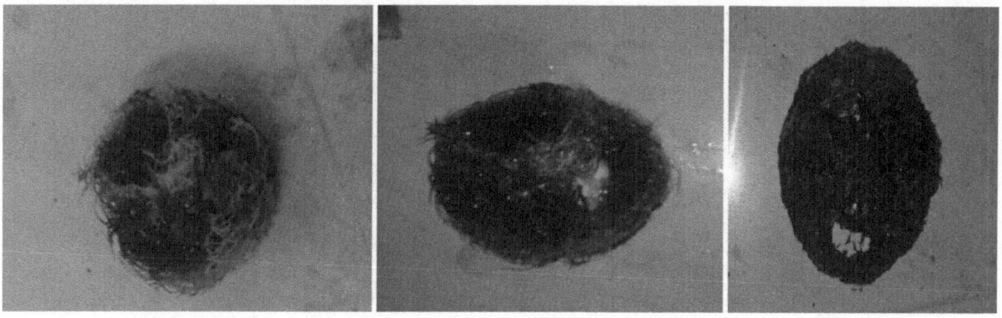

FIGURE 4.5 Biofilm generation of eco-friendly carrier (ridge gourd fibre) on the 7th, 10th, and 15th days.

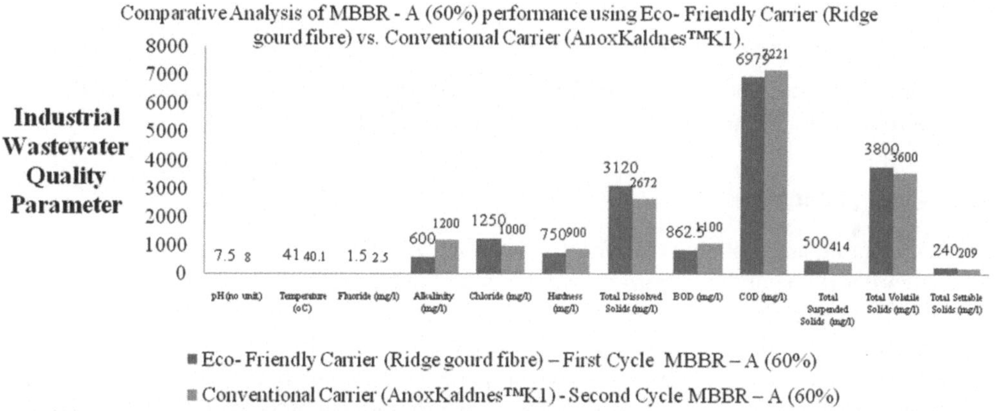

FIGURE 4.6 Comparative Analysis of MBBR – A (60%) performance using eco-friendly carrier (ridge gourd fibre) vs. conventional carrier (AnoxKaldnes™ K1).

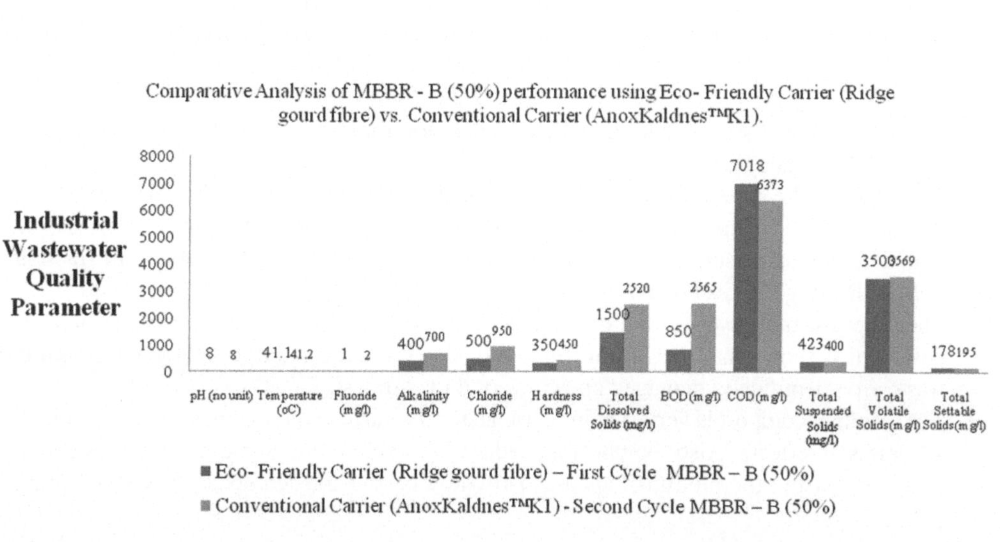

FIGURE 4.7 Comparative analysis of MBBR – B(50%) performance using an eco-friendly carrier (ridge gourd fibre) vs.a conventional carrier (AnoxKaldnes™ K1).

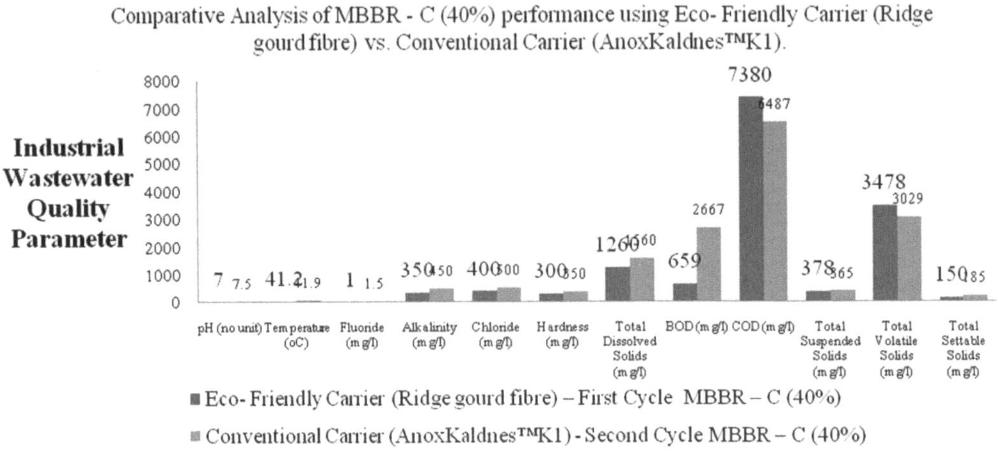

FIGURE 4.8 Comparative analysis of MBBR – C (40%) performance using an eco-friendly carrier (ridge gourd fibre) vs. a conventional carrier (AnoxKaldnes™ K1).

Case III: Comparative analysis of MBBR – C (40%) performance using an eco-friendly carrier (ridge gourd fibre) vs. a conventional carrier (AnoxKaldnes™ K1) in Figure 4.8.

4.6 CONCLUSION

The removal of organic pollutant parameters such as TDS, BOD, COD, and chlorides from industrial wastewater is one of the important issues in many industrialized countries. Lab-scale MBBR using eco-friendly carriers has gained reliability, as it offers a technically feasible and economical approach. In this regard, paper industry wastewater was collected by random sampling. The water samples were analyzed for various parameters as per Bureau of Indian Standards, before and after the wastewater treatment using the eco-friendly and conventional carriers for MBBR. The investigation continues with the design of the carrier with theoretical design. The purpose of selecting ridge gourd fibre as the eco-friendly carrier is because of its unique property of having pore size, and this factor plays a critical role in biofilm generation. Lab-scale MBBR has been designed and analyzed by varying the carrier fill of 40%, 50%, and 60% of the eco-friendly carrier (Ridge gourd fibre) and conventional carrier (AnoxKaldnes™ K1), respectively, for the inflow rate of 8L per day [12].

- The system was operated with a hydraulic retention time of 15 days to evaluate the lab-scale performance [12].
- According to the experimental results, the organic pollutant parameter removal efficiency was about 50% of the carrier fill of MBBR using an eco-friendly carrier, whereas the organic pollutant parameter removal efficiency was about 60% with a carrier fill fraction of MBBR using a conventional carrier.
- Then the research involved the disposal of biofilm-shaped eco-friendly carriers as compost, which shows that the harmful emissions are reduced if large volumes of organic waste are scientifically managed and converted to compost.
- The resultant compost is very rich in nutrients for the earth and replenishes topsoil without adverse side effects. Also, composting reduces the volume of waste that is thrown out. After converting the biofilm-shaped eco-friendly carrier into compost, it can be reused [12].

REFERENCES

American Public Health Association. (2005) *Standard Methods for Examination of Water and Waste Water*, 21st edition. Washington, DC: APHA.

Aygun, Ahmet, Nas, Bilgehan, and Berktay, Ali. (2008) "Influence of High Organic Loading Rates on COD Removal and Sludge Production in Moving Bed Biofilm Reactor", *Environmental Engineering Science*, Volume 25, Number 9, pages 28–32, © Mary Ann Liebert, Inc., doi:10.1089/ees.2007.0071.

Arneth, J.D., Midle, G., Kerndoff, H., Schleger, R. and Baccini, P. (1989) *Waste in Deposits Influence on Ground Water Quality as a Tool for Waste Type and Site Selection for Fnal Storage Quality*, Berlin: Springer Verlag, page 399.

Bhide, A.D. and Sundaresan, B.B. (1983). *Solid Waste Management in Developing Countries*, New Delhi, India: Indian National Scientific Documentation Center.

Borkar, R.P., Gulhane, M.L. and Kotangle, A.J. (2013) "Moving Bed Biofilm Reactor – A New Perspective in Wastewater Treatment", *IOSR-JESTFT*, Volume 6, Issue 6 pp. 15–21, e-ISSN: 2319-2402, p-ISSN: 2319-2399.

Bruins, M.R., Kepi, S. and O EHME, F.W. (2000) "Ecotox Environ Safe", *Power Point Presentation*, Volume 45, pages 198–207.

Casas, Monica Escola a, Chhetri, Ravi Kumar b, Ooi, Gordon a, Hansen, Kamilla M.S., b, Litty, Klaus c, Christensson, Magnus d, Kragelund, Caroline c, Andersen, Henrik R. b, and Bester, Kai a (2015) "Biodegradation of Pharmaceuticals in Hospital Wastewater by Staged Moving Bed Biofilm Reactors (MBBR)", *Water Research,* Volume 83, pp. 293–302, http://dx.doi.org/10.1016/j.watres.2015.06.042.

Chopra, Ashok Kumar and Sharma, Arun Kumar (2013) "Removal of Turbidity, COD and BOD from Secondarily Treated Sewage Water by Electrolytic Treatment", *Applied Water Science,* Volume 3, pages 125–132, doi: 10.1007/s13201-012-0066-x.

Environmental Engineering by G.N. Pandey & G.C. Carney, 13th edition, 2015 published by Tata McGraw - Hill Publishing Company Limited, New Delhi.

Kodavasal, Arneth S. (2014) *The STP Guide – Design, Operation & Maintenance*, KSPCB.

NPTEL IIT Kharagpur web courses – Module 18: Secondary Treatment Lecture 23.

Popa, Paula, Timofti, Mihaela, Voiculescu, Mirela, Dragan, Silvia, Trif, Catalin, and Georgescu, Lucian P. (2012) "Study of Physico-Chemical Characteristics of Wastewater in an Urban Agglomeration in Romania", *The Scientific World Journal,* Volume 12, pages 132–136, Article ID 549028, doi:10.1100/2012/549028.

Santhosh Kumar, R. (2016) "Assessment of Individual Household Waste Generation Quantity in Different Areas of Salem" published in *Bonfring International Journal.*

Santhosh Kumar, R. (2016) "Environmental Impact Assessment Disposal Site Pollution on Groundwater Quality of Salem", Presented in National Conference on Innovation in Concrete & Construction (ICON'16), conducted by Sona College of Technology, Salem on 30th March 2016 published in IJRIET.

Santhosh Kumar, R. (2016) "Environmental Impact Potential Assessment of Individual Household Waste Generation Quantity in of Salem", Presented and Published in International Conference (ICEIET-2016) conducted by Seventh Sense Research Group at Hindusthan Institute of Technology, Coimbatore on 18th & 19th March 2016 ICEIET- 2016 by SSRG.

Santhosh Kumar, R. (2017) "Contemporary Focuses on Integrated Solid Waste Management System through Small- Scale Onsite Composting for Educational Institution", Published in Sustainable Waste Management: Policies and Case Studies, 7th ICONSWM – ISWMAW 2017 – Volume – 1, Springer Nature.

Santhosh Kumar, R. (2017) "Contemporary Focuses on Integrated Solid Waste Management System through Small Scale Onsite Composting for Educational Institution & Integrated Approach on Prefeasibility Report of Biomedical Waste Management System", Presented in All India Seminar on "An Integrated Management Approach to Municipal Solid Wastes in India conducted by IEI (Salem Local Centre) at Sona College of Technology, Salem on 25th & 26th April 2017.

Santhosh Kumar, R. (2017) "Depletion of Ground Water Quality in Disposal Sites of Salem, Tamilnadu", Presented in International Conference on "Mechanics Materials & Management" conducted by College of Engineering, Trivandrum on 13th, 14th& 15th July 2017.

Santhosh Kumar, R. (2017) "Developing Integrated Organic Waste Management Plan", Presented in International Conference on Emerging Trends in Biotechnology for Waste Conversion (ETBWC-2017) 14th Convention of the Biotech Research Society, India (www.brsi.in) hosted by CSIR-National Environmental Engineering Research Institute, Nagpur, 8–10 October 2017.

Santhosh Kumar, R. (2017) "Evaluation of Household Waste Disposal Strategy in Different Areas of Salem", Presented in International Conference on Civil Engineering & Infrastructural issues in Emerging Economies (ICCIEE-2017) conducted by DST at SASTRA University, Thanjavur on 17th & 18th March 2017.

Santhosh Kumar, R. (2017) "Household Waste Disposal Strategy: Centralized vs. Decentralized Areas of Salem", Presented in International Conference on "Integrated Solid Waste Management Practices in Developing Countries" conducted by (CSIR-NEERI), Nagpur on 11th & 12th April 2017.

Santhosh Kumar, R. (2018) "Process Design & Experimental Analysis of Lab scale Moving Bed Biofilm Reactor using Eco-friendly Carrier", Presented in International Conference on Global Environmental Changes and Sustainable Development (ICGECSD 2018) conducted by Vel Tech Rangarajan Dr. Sagunthala R&D Institute of Science & Technology (Deemed to be University) on 4th & 5th of April 2018, published in IJARTET, 2018.

Santhosh Kumar, R. (2019) "Developing Integrated Rainwater Harvesting Plan for an Educational Institution at Salem, Tamilnadu", Presented in International Conference on Environmental Degradation: Key Challenges to Sustainable Development (on the occasion of Centenary Celebrations of IEI) held from 22nd to 24th November 2019 at the Institutions of Engineers (India), Telangana State Centre," Visvesvaraya Bhavan", Khariatabad, Hyderabad, Telangana, India.

Sewage Disposal and Air Pollution Engineering (Environmental Engineering – Vol. II), by Santhosh kumar Garg, 13th edition, 2013 published by Khanna Publishers.

Siti Nur Hatika, Abu Bakar et al. (2020) "Performance of a Laboratory-scale Moving Bed Biofilm Reactor (MBBR) and Its Microbial Diversity in Palm Oil Mill Effluent (POME) Treatment", *Process Safety and Environmental Protection*, Volume 142, October 2020, pages 325–335.

Wastewater Engineering Treatment and Reuse by Metcalf & Eddy, 4th edition, 2016 published by McGraw Hill Education (India) Private Limited, New Delhi.

Water Supply & Sanitary Engineering by G.S. Birdie & J.S. Birdie, 8th edition, 2013 published by Dhanpat Rai Publishing Company (P) Limited, New Delhi.

Wisam Sabeeh Al-Rekabi (2015) "Mechanisms of Nutrient Removal in Moving Bed Biofilm Reactors", *International Journal of Scientific & Engineering Research*, Volume 6, Issue 1, 497, pages 32–36, ISSN 2229-5518.

Yonghong, Liu et al. (2020) "Pilot Study on the Upgrading Configuration of UASB-MBBR with Two Carriers: Treatment Effect, Sludge Reduction and Functional Microbial Identification", *Process Biochemistry*, Volume 99, December 2020, pages 211–221.

5 Antimicrobial Agents in Germicidal Products

A Threat to the Environment and Human Health

Vaishnavi Shedge and Seema Sambrani

5.1 INTRODUCTION

Humans, plants, animals, and microorganisms are all inherent parts of an ecosystem. These microorganisms can be bacteria or viruses and can be dangerous at times. The outbreak of such infectious organisms has resulted in many pandemics and epidemics as have been reported in different parts of the world for centuries. The Coronavirus disease 2019 (COVID-19) was declared a global pandemic by the World Health Organization (WHO), and over 5.1 million cases have been reported globally to date. The COVID-19 infection is caused by severe acute respiratory syndrome coronavirus 2 (SARS-CoV-2) and has deeply impacted human health. The guidelines issued by the WHO to combat COVID-19 include maintaining social distancing, wearing a mask, and covering the mouth while coughing or sneezing. Other imperative preventive measures include cleaning hands using sanitizer, disinfectants, and wipes to fight against COVID-19. The sudden awareness about hand hygiene and health has boosted the demand for disinfection products globally. Following the pandemic, the use of wipes has increased, stated Benno Dorer, CEO of the Clorox company (Naidu Richa, 2020).

Maintaining hand hygiene helps reduce the transmission of infectious diseases. Washing one's hands every time is not convenient; hence, the use of sanitizers, disinfectants, and wipes becomes a practical approach. Disinfection products are classified as chemical-based, herbal-based, and semi-herbal-based. Chemicals such as alcohol, quaternary ammonium compounds, hydrogen peroxide, silver, chlorine, iodine, chlorhexidine, and triclosan are commonly used as active ingredients in chemical-based products (Torabi & Zahra, 2022). Although the herbal-based product is claimed to be of plant origin, it also contains harsh chemicals that are harmful to both humans and the environment. Alcohol is a commonly used ingredient in disinfecting and sanitizing solutions because of its bactericidal and fungicidal activity. Humans are being exposed to chemicals that contain skin irritants, allergens, and surfactants. Furthermore, microbes are acquiring resistance to chemical agents used in disinfectants. As a result, such resistant microorganisms raise the risk of infection, as they are difficult to treat and control. The chemicals used in disinfection products end up in sewage, polluting water resources. As a result, exposure to such harmful chemicals harms aquatic ecosystems (Dhama et al., 2021). Unawareness of the ingredients in disinfection products, improper use, and overuse are all having disastrous effects on human health and the environment.

This chapter focuses on various preventive steps practised to avoid infection, products used to avoid infection, and their uses. Furthermore, chemicals used in disinfection products and their side effects on humans, animals, and the environment are also discussed in detail.

DOI: 10.1201/9781003268925-5

5.2 PREVENTIVE STEPS TO AVOID INFECTION: CLEANING, SANITIZING, AND DISINFECTING

Microorganisms are present in water, soil, the human body, and the surfaces of objects. Not all microorganisms are safe; some can be infectious too. Thus, it is essential to discover a way to avoid being infected by such harmful microorganisms. The three main approaches, namely cleaning, sanitizing, and disinfecting are practised to avoid infections. They differ in terms of the products used and their efficiency in cleaning or killing germs. Before employing detergents, sanitizers, or disinfectants, it is necessary to determine the best approach for reducing microbial burden.

5.2.1 CLEANING

Cleaning is the mechanical activity of brushing or scrubbing surfaces to eliminate pathogens and reduce the microbial burden that must be completed before applying disinfectant to any surface. Water and products such as detergents, soap, and microfibre cloths are used for cleaning purposes that help remove dust, debris, and other organic residues. The cleaning process only removes and reduces microorganisms but does not kill them. The protocol for efficient hand-washing is summarized in Figure 5.1. Washing hands with soap and water for 15–30 seconds is an effective practice to avoid the spread of infection and maintain hand hygiene.

The process of cleaning a hard surface should begin with the least contaminated region and work its way up to the most soiled or contaminated area (WHO Team, 2020). Detergents contain physically active elements that aid cleaning by emulsifying, penetrating, dispersing, foaming, and wetting the surface. The physically active elements of detergents include surfactants (ionic and non-ionic surfactants), chemical compounds (alkaline, acidic, water conditioner, oxidizers), fillers, and in some cases enzymatic compounds, corrosion inhibitors, glycol ethers, and butyl cellosolve

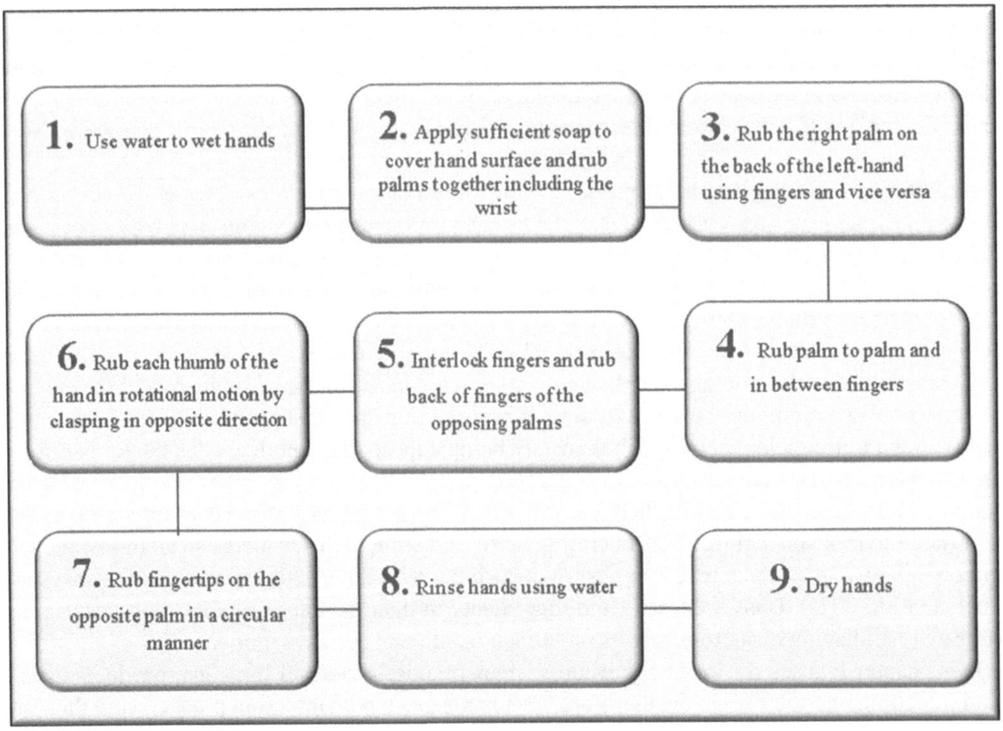

FIGURE 5.1 Protocol for efficient hand-washing (adapted from Tankeshwar, 2021).

that help remove oil, grease, and carbon (Schmidt, 2021). Frequently touched surfaces such as doorknobs, door push bars, sink handles, tabletops, and windows are expected to be cleaned regularly.

5.2.2 Sanitizing

The fundamental objective of the sanitizing method is to kill infectious microbes that may remain on the surface after cleaning and prevent their transmission. The sanitization process kills almost all microorganisms but not all. The chemical-based products used for sanitization are expected to be FDA- or EPA-approved, do not require potable water rinse, possess a wide range of activity, and have high stability, with less toxicity and corrosiveness. The sanitizer and surface must be kept in contact for the desired time as per the manufacturer's recommendation for removing microorganisms effectively. The sanitizers are considered safe for use on human skin (Minnesota Department of Health, 2017; Schmidt, 2021). The surfaces where food is served, stored, and prepared, hands, laptops, and computer screens, mobile phones, elevator buttons, etc., are expected to be sanitized regularly. Figure 5.2 provides instructions for proper hand sanitization.

5.2.3 Disinfecting

Disinfectant, according to *Merriam-Webster*, is an antiseptic solution consisting of chemical compounds that can kill infectious organisms present on non-living surfaces (Merriam-Webster, n.d.). The disinfection process is meant to kill microorganisms on hard and non-porous surfaces when exposed for a specific period, compared to a specific period as compared to sanitizer. Wiping or rinsing off the disinfectant solution before the recommended contact time, using a diluted disinfectant solution, or skipping the cleaning step does not guarantee the total reduction of microorganisms. Disinfectants do not have any effect on dirt, soil, or dust. Areas such as hospital floors,

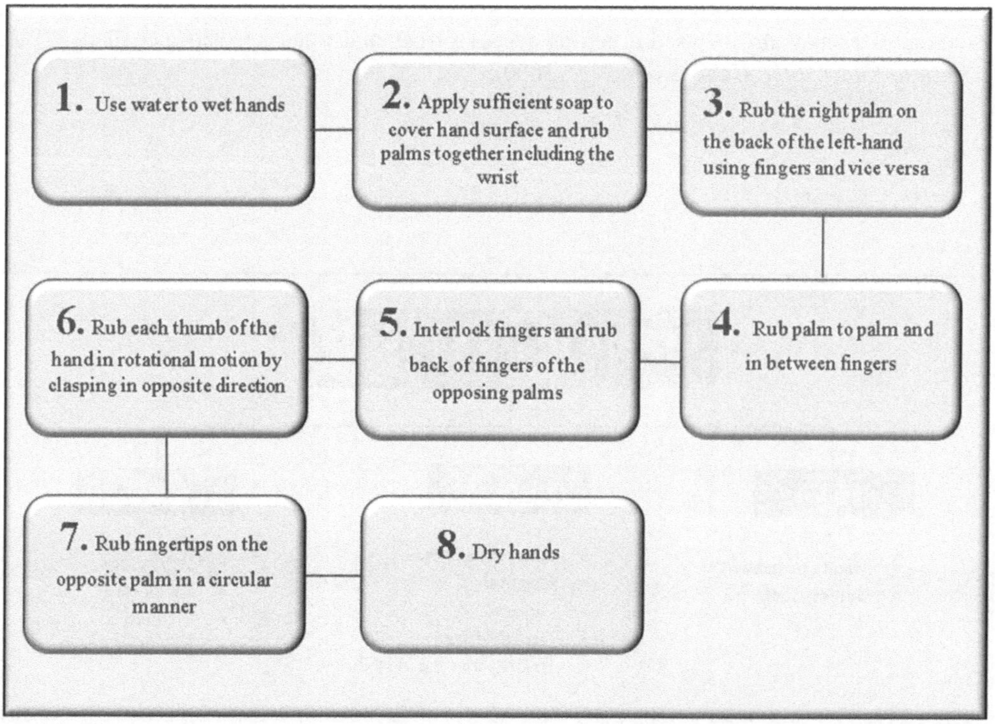

FIGURE 5.2 Protocol for efficient hand sanitization (adapted from Ontario, 2009).

bathroom sinks, toilet seats and floors, taps, doorknobs, and drinking fountains must be disinfected regularly to lower the risk of infections (Leonard et al., 2013).

5.3 PRODUCTS USED TO AVOID INFECTION AND ITS MULTIFARIOUS USES: SANITIZERS, DISINFECTANTS, AND WIPES

Amid COVID-19, a sudden hike in the use of germicidal products was seen. The virus or infectious bacteria may be present on surfaces, which are thus transmitted via hands, nose, mouth, etc. In order to prevent infection, various products are being used, *viz.*, sanitizers, disinfectants, and wipes. Figure 5.3 summarizes various products used for the prevention of infection and their varied uses.

5.3.1 Sanitizers

The WHO defines an alcohol-based hand rub as: "An alcohol-containing preparation (liquid, gel or foam) designed for application to the hands to inactivate microorganisms and/or temporarily suppress their growth" (Gold et al., 2022). Sanitizers are also known as antiseptics or hand rubs. They generally occur in the form of liquid, gel, or foam. Sanitizers are used on a larger scale in the healthcare sector. Healthcare workers use sanitizer when in contact with infectious patients, while performing antiseptic tasks, and, when in contact with bodily fluids or contaminated surfaces. The best option is always to use soap and water because this can kill all bacteria, including a few viruses that sanitizers can't kill. This includes a variety of actions, such as touching wet or dry waste at home, before consuming food, before, during, and after preparing food, after using the bathroom, after bandaging a cut or wound, and after changing soiled diapers. Additionally, it should be used before and after touching or treating a sick, diarrheic, or vomiting person, after handling animal faeces or products, and if your hands are dirty or oily. Apply hand sanitizer before and after using a public transit vehicle, while driving a vehicle (between breaks), when entering and exiting a public building, while inside and outside of retail stores, and before shaking hands or contact with another person. Sanitizer may also be used if getting access to soap and water presents a challenge (*What You Need to Know about Hand-Sanitizers?*, 2020).

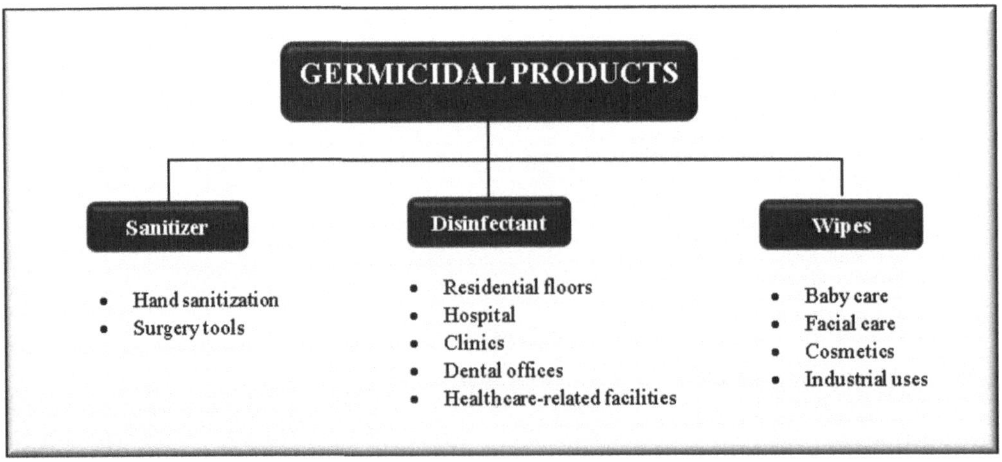

FIGURE 5.3 Types and uses of germicidal products.

5.3.2 Disinfectants

Disinfectant is a mixture of antimicrobial chemicals used to reduce microbial load from hard surfaces. Bodies such as the EPA and FDA approve disinfectants to be marketed in countries such as the United States. The EPA categorizes disinfectants into three classes: limited spectrum disinfectant, broad-spectrum disinfectant, and hospital disinfectant.

- **Limited spectrum disinfectant:** Particularly active against one type of microorganism. For example, a Gram-negative organism (*Salmonella enterica*) or a Gram-positive organism (*Staphylococcus aureus*).
- **Broad spectrum disinfectant:** Inhibitory effect against all microorganisms with a different Gram nature.
- **Hospital disinfectant:** A disinfectant with the properties of a broad-spectrum disinfectant and active against nosocomial pathogens such as *Pseudomonas aeruginosa.*

Limited and broad-spectrum disinfectants are generally used in households, institutes, or commercial sites. Hospital disinfectants, on the other hand, are used in healthcare facilities such as hospitals, dental clinics, etc. (Walleser Mitch, 2021).

5.3.3 Wipes

Wipes are used in day-to-day life for different purposes. The use of wet wipes has expanded in the past few decades for maintaining personal hygiene, household cleaning, baby care, cosmetics, and in many other industrial uses as well. The wipes are designed with their application in mind.

5.3.3.1 Baby Care

It has been reported that the cleaning solution in wet wipes for baby care includes water, mild detergents, moisturizing ingredients, perfume, and preservatives. Wet wipes are meant to effectively clean the delicate skin of the baby while keeping the conditions mild. The manufacturers of wet wipes claim that the chemicals employed in wet wipe production help avoid skin irritation and do not disturb the natural oils and beneficial microflora of the skin.

Wet wipes are known to be used for cleaning the face and hands post-meal. After a baby has defecated, wet wipes are often used to clean his or her genital area. Wet wipes with water as a major ingredient are used to clean the body of the baby. They are also used to clean the runny nose of children, spills, and to disinfect toys to reduce the risk of infection. Disinfecting wet wipes can also be used to clean the diaper disposal container and toilet training chairs of toddlers.

5.3.3.2 Facial Care and Cosmetics

The wipes containing cleaning solution can be used to clean the face, giving a feel of refreshment. They help to remove dirt and oil residues from the face, providing a feeling of freshness. Wet wipes with emollients are used to remove makeup from the face as they break down dirt, oil, and chemicals included in makeup. The facial care wipes improve wipe glide and hydrate skin, decreasing irritation and dryness. Wipes containing salicylic acid, vitamins, menthol, and other substances are used as anti-acne and anti-spot wipes (Crémieux et al., 2005; Xing et al., 2019).

5.3.3.3 Industrial Wipes

In the industrial sector, wet wipes are widely used to clean the workplace. Hard surfaces can be cleaned with wet wipes containing chemicals and solvents. Spunlace-based industrial wet wipes are durable and biodegradable, allowing pre-moistened chemicals to be discharged as needed. The construction and automobile industries make use of spunlace-based industrial wipes. Wet floor cleaning wipes are predicted to be in high demand shortly (Janmay Singh Hada, 2020).

5.4 CHEMICALS AND PRESERVATIVES USED IN MAKING DISINFECTION PRODUCTS AND THEIR SIDE EFFECTS ON HUMAN HEALTH

Disinfection products consist of a disinfecting solution, which is a mixture of various compounds. There is a range of products available on the market based on customer requirements. Apart from chemical-based products, herbal-based products are gaining a lot of attention due to their herbal ingredients.

Disinfecting products, such as wet wipes, contain water as the main ingredient; thus, they are prone to contamination. The addition of preservatives extends the shelf life of the product and prevents contamination. However, these preservatives are known allergens and carcinogens. Even though wipes are labelled as herbal-based wipes, they still contain a variety of chemicals and preservatives. As a result, there are serious concerns about the chemicals employed and their effects on human and animal health, as well as the negative consequences on the ecosystem.

Skin is the largest organ of the body and is hence subjected to a substantial amount of chemical exposure. During COVID-19, there was a surge in the usage of chemical-based disinfection products due to heightened awareness about hand cleanliness. Overuse of such products has resulted in several incidents of skin irritation, contact dermatitis, and allergies, according to experts. The aforementioned problems are seen in kids due to insufficient oil gland development (Dev, 2020). On 26 March 2021, a PubMed Central journal reported the hazards of cleaning with wipes containing disinfecting solutions: "The disinfecting cleaning wipes can burn". Exposure to harsh disinfectant harsh chemicals has increased by 25% since the COVID-19 pandemic. Dermatologists have reported a rise in painful skin conditions. They further stated that people should be mindful of the contents of sanitizing products (Dear et al., 2021). The overuse of alcohol-based sanitizers has also led to skin problems like dermatitis, dry skin, and alcohol poisoning (Froum, 2021). Chemicals such as alcohol, parabens, formaldehyde and formaldehyde releasers, isothiazolinones, and organic acids are the major class of chemicals incorporated in disinfecting products. Table 5.1 summarizes various chemicals used and their side effects. Though added with the intention to safeguard mankind, they are indirectly harming us.

5.4.1 Alcohol

Disinfection products contain alcohols such as ethanol, isopropanol, and n-propanol as active ingredients individually or in combination. Alcohol makes pathogens non-functional by denaturing or coagulating their proteins. Ethanol is the most commonly used ingredient in sanitizers and wet wipes as it shows broad-spectrum effectiveness against viruses. However, ethanol is known to induce dryness, irritation, and burning when applied at high concentrations. It induces lipid solubilization, fluidization, and protein denaturation of the stratum corneum. Alcohols like isopropanol cause yellow-brown discolouration and protein cross-linking (Goh et al., 2021). Phenoxyethanol, the most often used preservative in wet wipes, has a less damaging effect than other preservatives. Cases of skin irritation, hypersensitivity, and eye discomfort are less common when using wet wipes containing phenoxyethanol. Phenoxyethanol-containing products can induce contact urticaria and contact dermatitis; however, only a few cases have been reported (Dréno et al., 2019). Below are a few cases reported due to poisoning by alcohol-based disinfecting products.

Case study: A 13-year-old developed minor oedema with several superficial cracks, multiple fissures, and a small amount of blood on the dorsa and between several fingers due to continuous handwashing and concomitant application of alcohol-based hand sanitizer. The patient was later diagnosed with cumulative irritant contact dermatitis, a skin condition caused by chemicals or other physical agents damaging the stratum corneum before it can heal itself (Pope & Ousley, 2020).

Case study: Another case was reported wherein a 5-year-old was diagnosed with conjunctival congestion and superficial punctate keratopathy in the inferior quadrant of the cornea due to

TABLE 5.1
Chemical Used in Germicides and Their Side Effects

Chemical Category	Chemical Used	Side Effects
Parabens	Methylparaben, propylparaben, ethylparaben, butylparaben, benzylparaben	Reduction in sperm production, fertility issues, breast cancer and melanomas, endocrine disruptor, allergic reactions such as anogenital dermatitis.
Formaldehyde and formaldehyde donors	Paraformaldehyde, bronopol, quarternium-15, Imidazolidinyl urea, diazolidinyl urea, methenamine, poly(*p*-toluenesulfonamide coformaldehyde), dimethyloldimethyl (DMDM) hydantoin	Contact allergies, dermatitis and skin irritation, concomitant allergic reactions, asthma-like symptoms, and respiratory issues when inhaled, eye irritation, urinary tract irritation, digestive problems, and severe nephritis, harm aquatic life
Isothiazolinones	Methylchloroisothiazolinone methylisothiazolinone	Contact dermatitis, toxic to the human respiratory system, cytotoxic effect on the human liver and neuronal cell lines, carcinogen, skin irritation, skin sensitization.
Alcohols	Ethanol, isopropanol, phenoxyethanol	Dryness, irritation, burning, lipid solubilization, fluidization, and protein denaturation of the stratum corneum, yellow-brown discoloration and protein cross-linking, contact urticaria, and contact dermatitis.
Organic acids	Citric acid, glycolic acid, lactic acid	Burning, dermatitis, skin peeling, itching, and moderate sunburns upon dermal penetration, trigger the production of proinflammatory cytokines, including tumour necrosis factor and interleukin 1, epidermal and dermal thickening, reversible skin photosensitivity, contact urticaria, and keratinocyte apoptosis, decrease melanin levels in the skin.

alcohol-based sanitizer. Ethanol is a highly toxic chemical for corneal limbal stem cells and can cause a rapid cytotoxic effect in corneal epithelial cells (Yangzes et al., 2021).

Case study: A 12-year-old boy developed a skin rash on the dorsal side of his hands and experienced itching, dryness, and redness. Upon investigation, parents revealed a history of frequent hand sanitization by the child at least 9–10 times a day using a 70% isopropyl alcohol sanitizer. Later, the child was diagnosed with "isopropyl alcohol-based sanitizer-induced irritant contact dermatitis" due to repetitive use of the sanitizer. When the case was assessed with the Naranjo causality assessment score, the score came out to be 6, which indicated an adverse effect due to the use sanitizer containing 70% isopropyl alcohol (Inder & Kumar, 2020).

A case was reported in New Zealand, where a 3-year-old boy faced muscle damage and vision impairment due to chemical burns while using a hand sanitizer foot pedal dispenser at a restaurant. The incident of spraying the sanitizer into his eyes caused corneal epithelial damage, vision impairment, and a droopy eye (Stuff, 2020).

FDA and US poison control centres have reported side effects of alcohol-based sanitizer vapours, which include symptoms such as headache, nausea, and dizziness. A sudden spike in cases was observed amid COVID-19, which included side effects caused by dermal and nasal exposure to sanitizer (US Food and Drug Administration, 2021).

5.4.2 Parabens

Parabens are alkyl esters derived from *p*-hydroxybenzoic acid. Parabens are a user-friendly additive because they are stable, water-soluble, and antibacterial. Parabens are absorbed through the skin of humans. Upon exposure to parabens, they enter the systemic circulation and undergo rapid biotransformation in the liver. Parabens can cause a reduction in sperm production, fertility issues, breast cancer, and melanomas. They act as endocrine disruptors because they can interfere with the proper functioning of hormones involved in tissue development from fertilized sperm and eggs. Wet wipes containing parabens can cause allergic reactions including anogenital dermatitis. Methylparaben and propylparaben are the most widely used parabens in the manufacturing of wet wipes, followed by ethyl-paraben, butyl-paraben, and benzyl-paraben (Rocha et al., 2020).

5.4.3 Formaldehyde and Formaldehyde Donors

Formaldehyde and formaldehyde preservatives possess strong bactericidal and fungicidal activity, thus making them a perfect choice for use in personal and cosmetic products. However, they are known to cause contact allergies, dermatitis, and skin irritation. The concomitant allergic reactions due to formaldehyde and formaldehyde releasers are prevalent in sensitive patients (Jairoun et al., 2020). Some examples of formaldehyde releasers are paraformaldehyde, bronopol, quaternium-15, imidazolidinyl urea, diazolidinyl urea, methenamine, and poly(p-toluene sulfonamide-co-formaldehyde), dimethyloldimethyl (DMDM) hydantoin (Elder & Crowley, 2012; Siti Zulaikha et al., 2015).

DMDM hydantoin, another chemical commonly used in wipes, is hazardous when swallowed, and it can cause asthma-like symptoms and respiratory issues when inhaled. It can also lead to eye irritation (National Center for Biotechnology Information, 2022a).

Case study: A case was reported in the *Journal of the American Academy of Dermatology*, wherein a 36-year-old woman developed pruritic blisters on her hands a few days after the delivery of her child. Upon examination, it was revealed that she had erythematous scaling patches with fissures involving the ventral and lateral surfaces of her fingers and palms. A biopsy specimen of the hand showed parakeratosis, spongiosis, dermal oedema, and collections of Langerhans cells in the epidermis. A patch test was performed with readings at reads of 48 and 72 hours. The patient showed positive reactions to formaldehyde, quaternium-15, nickel sulphate, diazolidinyl urea, DMDM hydantoin in a petroleum base, DMDM hydantoin in an aqueous solution, imidazolidinyl urea, MCI/MI, palladium chloride, and ethyleneurea melamine-formaldehyde. After investigation, it was found that wipes she used for infant diaper hygiene contained DMDM hydantoin. After proper medication, she was able to cope with the allergy (Blume et al., 2006).

Bronopol is a preservative widely used and can be hazardous if ingested. It can induce acute toxicity, dermal irritation, skin corrosion/irritation, severe eye injury, or eye irritation upon contact with the skin. When inhaled, it may cause respiratory tract infection. Also, bronopol has the potential to harm aquatic life (National Centre for Biotechnology Information, 2022b).

Methenamine is dangerous whether absorbed through the skin, inhaled, or consumed. Upon contact with the skin, it can cause a rash, redness, pain, and burns. It irritates the eyes, respiratory tract, skin, and mucous membranes. Shortness of breath, coughing, and inflammation of the respiratory tract can all be caused by inhaling methenamine. It can cause redness, pain, and blurred vision upon contact with the eyes. It can cause urinary tract irritation, digestive problems, and severe nephritis and is fatal if consumed. Repeated use of methenamine-containing products can cause skin sensitivity, urticaria, and dermatitis. Long-term exposure to such products might cause skin irritation and reddening (National Centre for Biotechnology Information, 2022c).

5.4.4 Isothiazolinones

Isothiazolinones, like methylchloroisothiazolinone (MCI) and methylisothiazolinone (MI), are found in wet wipes and cosmetics. There have been reports of contact dermatitis caused by inflammatory mediators in skin cells. Also, it can be toxic to the human respiratory system if inhaled. It can have a cytotoxic effect on the human liver and neuronal cell lines. It has the potential to cause cancer in humans (Silva et al., 2020). Chemical burns caused by MCI and MI irritate the skin and can cause skin sensitization when used in high concentrations. They primarily affect the eye and hand areas. Wet wipes containing MI harm children's perioral skin, vaginal area, and other genitals (Herman et al., 2019).

Case study: A case was reported in the *Thai Journal of Dermatology*, where a 28-year-old Thai woman had itchy rashes and lesions throughout her fingertips, finger webs, and palms on both hands. Physical examination suggested erythematous papules and deep-seated vesicles on the erythematous base and swelling of both hands. The survey taken at the workplace revealed she had skin lesions due to her occupation. The survey further revealed that the lesions were due to hand sanitizer. Later, the patient was subjected to a patch test at the Contact and Occupational Clinic, Institute of Dermatology in Bangkok. The patient tested positive for MCI/MI, MI, lanolin alcohol, and nickel sulphate hexahydrate. The diagnosis suggested that the patient was suffering from occupational allergic contact dermatitis due to MI-containing hand sanitizers (Prasan, 2021).

Case study: A case was reported in the *Contact Dermatitis Journal*, wherein a 48-year-old woman, a hairdresser by profession, showed hand dermatitis on the palms and fingers for a duration of three years. Upon patch testing, the patient tested positive for MCI/MI and indicated contact dermatitis. Later, it was discovered that she used wet wipes containing MCI/MI. The condition improved when she stopped using wet wipes (Timmermans et al., 2007).

5.4.5 Organic Acids

Organic acids such as citric acid, glycolic acid, and lactic acid can cause burning, dermatitis, skin peeling, itching, and moderate sunburns upon dermal penetration. When in contact with humans, they can trigger the production of proinflammatory cytokines, including tumour necrosis factor and interleukin 1. At higher concentrations, they lead to epidermal and dermal thickening. Reversible skin photosensitivity, contact urticaria, and keratinocyte apoptosis have all been reported, all of which exacerbate skin photodamage. They also decreases melanin levels in the skin (Goh et al., 2021).

5.5 ADVERSE EFFECT OF DISINFECTANT PRODUCT ON THE ECOSYSTEM

Demand for a variety of disinfection products has exploded since the onset of COVID-19 and has increased global awareness of sanitation and personal hygiene. Many disinfection product manufacturing companies have raised production to satisfy worldwide demand. Wildlife is extremely affected by the usage of such anthropogenic chemicals. Around 100 animals and birds were found dead in China, which included 17 species. The report noted that the species of common blackbird, *Turdus merula*, and Siberian weasel *Mustela sibirica*, were found dead in Chongqing, China, due to the overuse of disinfectants (Shalw, 2020). Ancona, a city in Italy, also reported cases of bird deaths due to street sanitization (Appignanesi, 2020).

Wipes are one of the biggest marine contaminants found around coastlines, beaches, and inside water bodies because of their overuse. Wet wipes that are non-degradable break down into microplastics and nanoplastics. Marine animals ingesting microplastics are thus entering food systems and posing a threat to human health.

The microfibres from wet wipes are reported to be a cause of synthetic microplastic pollutants in natural environments. Humans are exposed to these microplastic fibres during the cleaning process.

The improper disposal of wipes might result in the production of millions of microplastic fibres, thus making it mandatory to dispose of them properly (Ó Briain et al., 2020). Also, the amount of garbage dumped in landfills and incinerators is increasing on a daily basis due to the overuse of single-use PPE wet wipes. Marine systems are being polluted as microplastic from landfill leachates and incinerated waste is released through sewage sludge into water bodies. Flushing wipes into toilets has also led to blocked sewage systems. The packaging of wipes in plastic boxes or wraps adds to microplastic contamination if not discarded properly (Shruti et al., 2021).

The process of disinfecting exterior floors in apartments and hospitals, roadways, and marketplaces is additionally contributing to disinfectant flow into sewers, rivers, and lakes. The disinfectant containing sodium hypochlorite is used in healthcare settings to prevent nosocomial infections. If these streams are left untreated, they may enter water bodies, thus contaminating drinking water sources. The disinfectant containing chlorine reacts with dissolved organic matter in surface water to produce by-products such as haloacetic acids and trihalomethanes. The generation of organic chlorine compounds persists as environmental contaminants and may pose a considerable risk to aquatic ecosystems. Excessive use of disinfectants can lead to bioaccumulation, biomagnification, toxicity, as well as mutations, the spread of antibiotic resistance genes, and the development of antibiotic-resistant bacteria (Dhama et al., 2021).

The bacterial resistance to the chemicals used in disinfectants has been observed recently. Bacteria exhibit such resistance to disinfectants by two mechanisms. The first mechanism involves the outer layers of cells which prevent the fatal action of disinfectants by interrupting disinfectant molecules from reaching the target region in sufficient quantity. This mechanism is called intrinsic insusceptibility and is prevalent in spores and mycobacteria. Biofilm development and triclosan efflux, chlorhexidine efflux, and quaternary ammonium compounds efflux are some other examples of intrinsic insusceptibility mechanisms. An acquired mechanism, the second type of mechanism, leads to genetic alteration in the bacterial cell by mutation, selection, or acquisition of plasmids or transposons. These changes in genetic makeup contribute to a method to counteract the fatal effects of disinfectants. Both Gram-positive and Gram-negative bacteria have developed resistance to chemicals by acquired mechanisms. These bacteria, like antibiotic-resistant bacteria, can transmit infections and become more difficult to treat (Russell, 2002).

5.6 CONCLUSION

Maintaining hand hygiene has become more vital than it has ever been. Various processes are being practised by us to keep our families and loved ones safe and healthy. Cleaning, disinfecting, and sanitizing have become the new normal. Various products such as wet wipes, sanitizers, and disinfectants are used by healthcare staff, patients, and the general community to combat the novel Coronavirus disease. The significant rise in the use of disinfection products has raised concerns too. Although the disinfection product brands in the market claim to be herbal and harmless, the chemicals included during production give rise to alarming health conditions. Chemical-based products have a negative influence on human health, animals, microorganisms, and the environment. The herbal-based products, though branded as plant-origin, contain several chemicals that could be harmful. This chapter covered various preventive steps practised to avoid infection, products used to avoid infection and their uses, chemicals used in disinfection products and their side effects on humans, animals, and the environment. The overuse of disinfectants, as well as a lack of understanding about their use, has raised concerns about resistant bacteria and microplastic pollution. Keeping in mind the current pandemic scenario, it is mandatory to be armed with knowledge about the strategies to and products employed to maintain hygiene and comfort.

REFERENCES

Appignanesi, A. (2020, April 16). Disinfections against Covid and bird deaths, Brutti of the Enpa: “Especially birds at risk.” *Centropagina*. https://www.centropagina.it/attualita/ancona-disinfezioni-anti-covid-morie-uccelli-enpa-avifauna/

Blume, J. E., Stoll, H. L., & Cheney, R. T. (2006). Treatment of primary cutaneous CD30 + anaplastic large cell lymphoma with intralesional methotrexate. *Journal of the American Academy of Dermatology, 54*(5 Suppl. 1), 230–232. https://doi.org/10.1016/j.jaad.2005.11.1037

Crémieux, A., Cupferman, S., & Lens, C. (2005). Method for evaluation of the efficacy of antimicrobial preservatives in cosmetic wet wipes. *International Journal of Cosmetic Science, 27*(4), 223–236. https://doi.org/10.1111/j.1467-2494.2005.00267.x

de Mélo Silva, I. S., do Amorim Costa Gaspar, L. M., Rocha, A. M. O., da Costa, L. P., Tada, D. B., Franceschi, E., & Padilha, F. F. (2020). Encapsulation of red propolis in polymer nanoparticles for the destruction of pathogenic biofilms. *AAPS PharmSciTech, 21*(2). https://doi.org/10.1208/s12249-019-1576-8

Dear, K., Nguyen, J., & Nixon, R. (2021). Disinfectant cleaning wipes can burn! The hazards of cleaning without adequate skin protection. *Contact Dermatitis, 85*(1), 109–110. https://doi.org/10.1111/cod.13800

Dev, A. (2020, June 22). Delhi-NCR: Rashes or eczema? It’s probably your sanitiser | Noida News - Times of India. *The Times of India* . https://timesofindia.indiatimes.com/city/noida/rashes-or-eczema-its-probably-your-sanitiser/articleshow/76500291.cms

Dhama, K., Patel, S. K., Kumar, R., Masand, R., Rana, J., Yatoo, M. I., Tiwari, R., Sharun, K., Mohapatra, R. K., Natesan, S., Dhawan, M., Ahmad, T., Emran, T. Bin, Malik, Y. S., & Harapan, H. (2021). The role of disinfectants and sanitizers during COVID-19 pandemic: Advantages and deleterious effects on humans and the environment. *Environmental Science and Pollution Research, 28*(26), 34211–34228. https://doi.org/10.1007/s11356-021-14429-w

Dréno, B., Zuberbier, T., Gelmetti, C., Gontijo, G., & Marinovich, M. (2019). Safety review of phenoxyethanol when used as a preservative in cosmetics. *Journal of the European Academy of Dermatology and Venereology, 33*(S7), 15–24. https://doi.org/10.1111/jdv.15944

Elder, D., & Crowley, P. (2012, January 1). *Antimicrobial preservatives part three: Challenges facing preservative systems*. American Pharmaceutical Review – The Review of American Pharmaceutical Business & Technology. https://www.americanpharmaceuticalreview.com/Featured-Articles/38874-Antimicrobial-Preservatives-Part-Three-Challenges-Facing-Preservative-Systems/

Froum, S. (2021, July 6). The dangers of hand sanitizer use and misuse | Perio Implant Advisory. *Perio-Implant Advisory*. https://www.perioimplantadvisory.com/clinical-tips/article/14206274/the-dangers-of-hand-sanitizer-use-and-misuse

Goh, C. F., Chiau, M. L., & Ching, W. L. (2021). Dermatologic reactions to disinfectant use during the COVID-19 pandemic. *Clinics in Dermatology, 39*, 314–322. https://doi.org/10.1016/j.clindermatol.2020.09.005

Gold, N. A., Mirza, T. M., & Avva, U. (2022). Alcohol sanitizer. *StatPearls*. https://www.ncbi.nlm.nih.gov/books/NBK513254/

Herman, A., Aerts, O., de Montjoye, L., Tromme, I., Goossens, A., & Baeck, M. (2019). Isothiazolinone derivatives and allergic contact dermatitis: A review and update. *Journal of the European Academy of Dermatology and Venereology, 33*(2), 267–276. https://doi.org/10.1111/jdv.15267

Inder, D., & Kumar, P. (2020). Isopropyl alcohol (70%)-based hand sanitizer-induced contact dermatitis: A case report amid Covid-19. *Indian Journal of Case Reports,* 6(7), 403–405. Doi 10.32677/IJCR.2020.v06.i07.017

Jairoun, A. A., Al-Hemyari, S. S., Shahwan, M., Zyoud, S. H., & Ashames, A. (2020). Hidden formaldehyde content in cosmeceuticals containing preservatives that release formaldehyde and their compliance behaviors: Bridging the gap between compliance and local regulation. *Cosmetics, 7*(4), 1–10. https://doi.org/10.3390/cosmetics7040093

Janmay Singh Hada. (2020). New trends in non-woven wet wipes. *International Journal for Modern Trends in Science and Technology, 06*(9S), 89–96. https://doi.org/10.46501/ijmtst0609s15

Leonard, V., Westinghouse, C., Bradman, A., Berns, J. E., Flattery, J., Jordan, D., Kunitz, J., Lipman, J., Adams, M., Seikel, K., Williams, B., Kempter, C., & Marshall, J. (2013). *Green cleaning, sanitizing, and disinfecting: A toolkit for early care and education*. UCSF Institute for Health & Aging, UC Berkeley Center for Environmental Research and Children’s Health, Informed Green Solutions, and California Department of Pesticide Regulation. www.joannagreeneditor.com

Merriam-Webster. (n.d.). *Disinfectant definition & meaning - merriam-webster*. Retrieved September 17, 2022, from https://www.merriam-webster.com/dictionary/disinfectant

Minnesota Department of Health. (2017). *Evaluation of cleaners, sanitizers, and disinfectants for surfaces*, 1–31. www.health.state.mn.us

Naidu Richa. (2020). Clorox won't have enough disinfecting wipes until 2021, its CEO says | Reuters. *Reuters*. https://www.reuters.com/article/us-health-coronavirus-clorox-wipes-idUSKCN2501EU

National Center for Biotechnology Information. (2022a). *1,3-Bis(hydroxymethyl)-5,5-dimethylimidazolidine-2,4-dione | C7H12N2O4 - PubChem*. UbChem Compound Summary for CID 22947. https://pubchem.ncbi.nlm.nih.gov/compound/DMDM-Hydantoin

National Center for Biotechnology Information. (2022b). *Bronopol | C3H6O4BrN - PubChem*. PubChem Compound Summary for CID 2450, Bronopol. https://pubchem.ncbi.nlm.nih.gov/compound/Bronopol

National Center for Biotechnology Information. (2022c). *Methenamine | C6H12N4 - PubChem*. PubChem Compound Summary for CID 4101, Methenamine. https://pubchem.ncbi.nlm.nih.gov/compound/Methenamine

Ó Briain, O., Marques Mendes, A. R., McCarron, S., Healy, M. G., & Morrison, L. (2020). The role of wet wipes and sanitary towels as a source of white microplastic fibres in the marine environment. *Water Research*, *182*, 116021. https://doi.org/10.1016/j.watres.2020.116021

Ontario, P. H. (2009, October). How to hand rub. *Queen's Printer for Ontario, Catalogue*. https://www.publichealthontario.ca/-/media/Documents/J/2009/jcyh handrub.pdf?rev=87165b8a22cc41948bb49a3128743906&sc_lang=en

Pope, V., & Ousley, L. (2020). Irritant contact dermatitis caused by hand sanitizer useand handwashing during the covid-19 pandemic. *Consultant*, *60*(7), 1–3. https://doi.org/10.25270/con.2020.04.00015

Prasan, S. (2021). Occupational contact dermatitis caused by methylisothiazolinone: A hidden culprit in the workplace. *Thai Journal of Dermatology*, *37*(4), 161–166.

Rocha, B. A., Bocato, M. Z., Latorraca, E. F., Ximeneza, J. P. B., & Barbosa, F. (2020). A survey of parabens in commercial baby wipes from Brazil and estimation of daily exposure. *Quimica Nova*, *43*(4), 442–446. https://doi.org/10.21577/0100-4042.20170494

Russell, A. (2002). Bacterial resistance to disinfectants. *British Journal of Infection Control*, *3*(3), 22–24. https://doi.org/10.1177/175717740200300306

Schmidt, R. H. (2021). *Basic elements of equipment cleaning and sanitizing in food processing and handling operations*. IFAS Extension, University of Florida, 1–11.

Shalw. (2020, February 21). *More than 100 wild animals drop dead near coronavirus epicentre in China - Sound health and lasting wealth*. Sound Health and Lasting Wealth. https://www.soundhealthandlastingwealth.com/uncategorized/more-than-100-wild-animals-drop-dead-near-coronavirus-epicentre-in-china/

Shruti, V. C., Pérez-Guevara, F., & Kutralam-Muniasamy, G. (2021, June). Wet wipes contribution to microfiber contamination under COVID-19 era: An important but overlooked problem. *Environmental Challenges*, *5*, 100267. https://doi.org/10.1016/j.envc.2021.100267

Siti Zulaikha, R., Sharifah Norkhadijah, S. I., & Praveena, S. M. (2015). Hazardous ingredients in cosmetics and personal care products and health concern: A review. *Public Health Research*, *5*(1), 7–15. https://doi.org/10.5923/j.phr.20150501.02

Stuff. (2020, December 23). New Zealand news: Young boy receives chemical burn on the eye from hand sanitiser at Wellington restaurant. *9News*. https://www.9news.com.au/health/new-zealand-news-young-boy-receives-chemical-burn-on-the-eye-from-hand-sanitiser-at-restaurant/5d17cc56-9776-49b7-a2cc-10132d8a8b8b

Tankeshwar, A. (2021). *Handwashing and hygiene: Why do we wash our hands – Microbe online*. Microbiome Online - Bacteriology. https://microbeonline.com/handwashing-hygiene-wash-hands/

Timmermans, A., De Hertog, S., Gladys, K., Vanacker, H., & Goossens, A. (2007). "Dermatologically tested" baby toilet tissues: A cause of allergic contact dermatitis in adults. *Contact Dermatitis*, *57*(2), 97–99. https://doi.org/10.1111/j.1600-0536.2007.01161.x

Torabi, S., & Zahra, F. (2022). Disinfectants. *StatPearls*. https://www.ncbi.nlm.nih.gov/books/NBK574540/

U.S. Food And Drug Administration. (2021). *FDA Warns that vapors from alcohol-based hand sanitizers can have side effects*. https://www.fda.gov/drugs/fda-drug-safety-podcasts/fda-warns-vapors-alcohol-based-hand-sanitizers-can-have-side-effects

Walleser Mitch. (2021, April 21). *Types of cleaners | Disinfectants for industrial, hospitals, or medical devices. How are they used?* Production Automation Corporation. https://blog.gotopac.com/2020/04/21/difference-between-a-sanitizer-vs-disinfectant-industrial-medical/

What you need to know about hand-sanitizers? (2020, December 10). Kauvery Hospital. https://kauveryhospital.com/blog/lifestyle/what-you-need-to-know-about-hand-sanitizers/

WHO Team. (2020, May 15). *Cleaning and disinfection of environmental surfaces in the context of COVID-19.* WHO Interim Guidance. https://www.who.int/publications/i/item/cleaning-and-disinfection-of-environmental-surfaces-inthe-context-of-covid-19

Xing, H., Krogmann, A. R., Vaught, C., & Chambers, E. (2019). Understanding the global sensory landscape for facial cleansing/makeup remover wipes. *Cosmetics*, *6*(3), 1–12. https://doi.org/10.3390/COSMETICS6030044

Yangzes, S., Sartaj, G., Gailson, T., & Grewal, S. P. S. (2021). Hand sanitizer–induced ocular injury: ACOVID-19 hazard in children. *JAMA Ophthalmology*, *193*(3), 362–364. https://doi.org/10.1007/s00417-007-0758-3

6 Role of Environment in Antibiotic Resistance Development

Anchal Lodhi, Deepak Kumar Arya, and Nidhi V. Maheshwari

6.1 INTRODUCTION

An important advancement in medical science was made in 1929 when Alexander Fleming discovered penicillin, which is often described as one of the important breakthroughs in medical science. Penicillin, with its ability to inhibit bacterial infections, particularly those caused by *Staphylococcus aureus*, transformed the way that many diseases were treated and cleared the path for the discovery of more antibiotics. Antibiotics are essential for both preventing and treating infections in people with chronic illnesses such as rheumatoid arthritis, tuberculosis, renal disease, and diabetes as well as those who have had significant surgeries like heart surgery or organ transplants. Urine and excrement of both animals and humans are contaminating the environment with antibiotics.

Antibiotics are typically categorized as "broad-spectrum", meaning they attack multiple bacteria, or "narrow-spectrum", suggesting they are dangerous to only a few microorganisms.

A variety of strategies are used by antibiotics to combat infections (Grenni et al., 2018):

1. Alteration of cell membranes
2. Disruption of the synthesis and repair of cell walls
3. Disruption of translation (50S rRNA and 30S rRNA)
4. Termination of replication
5. Antagonistic and anti-metabolite traits

Due to the widespread development of antibiotic resistance, antibiotics have become a "threatened species" due to their rapidly going extinct or degenerating (Founou et al., 2016).

Agents classified as antimicrobials either eradicate or stop the growth of bacteria. Antibiotics, insecticides, antifungals, and antivirals are examples of such antimicrobials. Microorganisms that were previously sensitive to antibiotic treatment develop resistance to it, which is known as antimicrobial resistance. Antimicrobial resistance is a result of the widespread usage of antibiotics, which reached 63,151 tonnes in 2010 and is expected to rise by 67% in 2030 (Kraemer et al., 2019). The environment is acknowledged by the World Health Organisation (WHO) as a major contributor to the development of antibiotic resistance (Bondarczuk et al., 2015; Palmer et al., 2018). Antibiotic-resistant bacterial infections have a significant monetary impact; in the European Union, they cost EUR 1.5 billion annually in healthcare expenditures and lost productivity, while in the US they cost $55 billion (Mobarki et al., 2019). Antimicrobial resistance can spread in the environment due to vertical and horizontal gene transfers via recombination and mutation. Environmental microorganisms play a significant role in providing forecast data regarding the emergence of antibiotic

DOI: 10.1201/9781003268925-6

resistance in the environment (Zhang et al., 2018). In the environment, the agriculture sector, comprising of crops, aquaculture and fisheries, livestock and companion animals, and forestry and wildlife, utilizes antimicrobials to prevent the diseases. In the industry, antimicrobials are used in food production and packaging, sanitizers, textile manufacturing, paints, coatings, additives, preservatives, and petroleum recovery. Antimicrobial residues from agriculture and industries contaminate the ecosystems mainly through solid waste and effluents.

6.2 ANTIMICROBIAL RESISTANCE (AMR)

AMR is a natural process, involving a genetic mutation in the microorganisms that significantly alters an aspect of their anatomy or physiology, allowing them to evade or resist the action of antimicrobial activity. AMR has posed a significant cost to society in terms of increased misuse/overuse, mortality, and morbidity.

An estimated 4.95 million fatalities globally in 2019 were linked to antimicrobial resistance (AMR), with 1.27 million of those being directly linked to resistant bacteria (Murray et al., 2022). The resistant strain can survive in the presence of antibiotics, whereas other strains without this mutation cannot survive. Infected hosts, *viz.* human, animal, or plant, can spread their AMR strains by direct and indirect transmission (Figure 6.1), such as via food and the environment; human and animal contact. For example, as a result of antibiotic resistance, tuberculosis is transmitted between people through the air, claiming the lives of 1.6 million people in 2021 after 10 million cases and 1.5 million related fatalities in 2020 (WHO, 2021). As per the report of the Indian Council of Medical Research (ICMR), *Klebsiella pneumoniae*'s sensitivity to meropenem decreased from 15% to 2.5% from 2018 to 2022, respectively, and for sensitivity to colistin it decreased from 96% to 28% from 2018 to 2022, respectively (Sharma et al., 2023).

6.3 FACTORS CONTRIBUTING THE EMERGENCE OF AMR

Generally, there are a few industries that are most affected by AMR susceptibility and dissemination, such as agriculture and animal production, human medicine in healthcare, and the environment sector (Figure 6.2). Antibiotics exert selective pressure on sensitive bacteria, killing or inhibiting them, while bacteria that are resistant by nature or that have developed resistance to antibiotics are more likely to survive and proliferate. The genesis and spread of high antibiotic resistance in hospitals is likely primarily caused by the extensive and continuous use of antimicrobial medications.

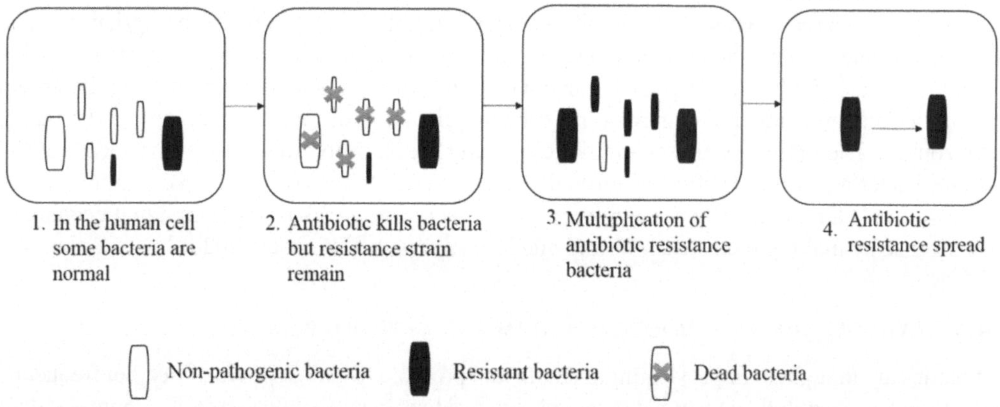

FIGURE 6.1 The transmission of antibiotic-resistant bacteria through air in humans.

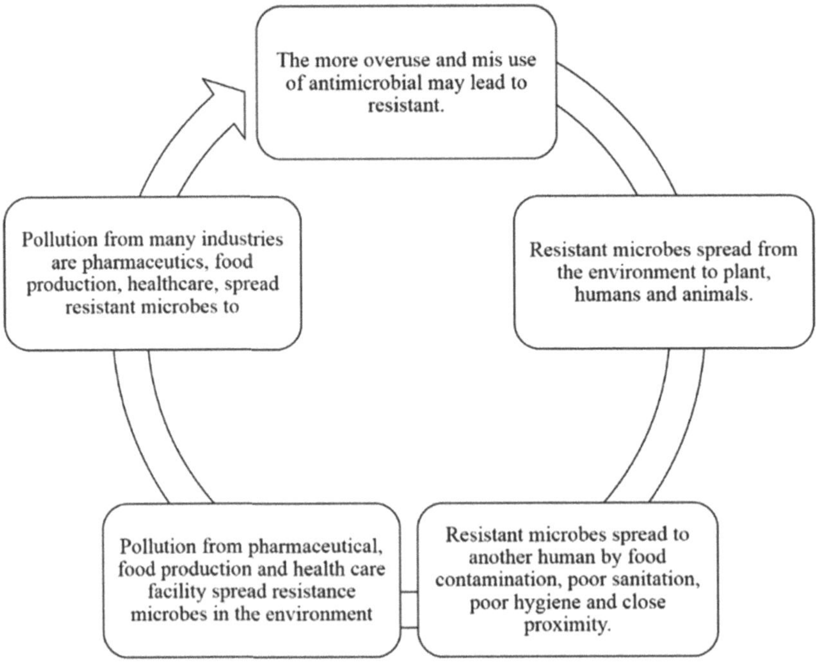

FIGURE 6.2 Spread of AMR in the environment.

6.4 CAUSES OF ANTIBIOTIC RESISTANCE

6.4.1 Incorrect Antibiotic Prescription

The Center for Disease Control (CDC) reports that improper antibiotic selection and duration are seen in 30–50% of clinical cases and treatment plans. Patients may experience therapeutic effects and clinical complications as a result of an incorrectly given antibiotic (Luyt et al., 2014). Antibiotic-induced changes in gene expression can initiate pathogenicity, whereas enhanced instability and horizontal gene transfer can facilitate the development of antibiotic resistance (Viswanathan, 2014).

6.4.2 Overuse/Misuse of Antibiotics

Antibiotics are generally used to treat diseases caused by bacteria in animals, humans, and plants. The majority of antibiotics ingested by humans (50–70%) are β-lactam antibiotics, which include cephalosporins, carbapenems, and penicillins. These antibiotics are used in the agriculture sector and medicine for preventive, metaphylactic, and growth-promoting purposes (Kumar and Pal, 2018). In contrast, antibiotics can be used in a wide range of ways, including promoting the growth of fruit, beekeeping, and breeding, as well as the amplification of the meat's protein and reducing its fat content. Therefore, overuse of antibiotics in daily life leads to drug-sensitive bacteria, causing resistant bacteria to evolve naturally through selection (Read and Woods, 2014). Antibiotics are overprescribed globally despite cautions regarding misuse and a lack of regulation that encourages their overuse by making them easier to find and less expensive (Iwu et al., 2020).

6.4.3 Extensive Use of Antibiotic in Animal and Agriculture Sector

Antibiotics are thought to improve animal health and promote livestock productivity, but frequently, these drugs are eventually consumed by people via residues in animal flesh, which suppress vulnerable bacteria and foster the growth of resistant strains (Golkar et al., 2014). Antibiotic remnants in animal products are caused by a number of factors, including physicochemical features, drug pharmacokinetics, and biological processes in the animal. These factors are linked to the inappropriate use of antibiotics. For more than 35 years, animal products have been used to transmit antibiotic resistance genes to people, particularly in areas where farmers have genes that are similar to those found in the intestinal flora of farm animals (Bartlett et al., 2013).

In the agriculture area, there are multifarious uses of antibiotics that affect the environmental microbiome. The majority of antibiotics given to cattle are excreted in urine and faeces, which spread into surface runoff, groundwater, manure, and other places, eventually polluting soil, water, and air (Zhang et al., 2018). The overuse of antibiotics in animal farming leads to agriculture ecosystem contamination by using contaminated manure in the agriculture sector. Antibiotic residues have a detrimental impact on the biota in the environment at various trophic levels through the use of contaminated food and water, as well as by contributing to the growth of the population of resistant bacteria and sustaining the selective pressure that leads to the emergence or spread of resistance in various environmental compartments.

6.5 ANTIBIOTIC RESIDUES AND THEIR EFFECTS ON THE ENVIRONMENT

Antibiotics are heterogeneous compounds considered as "persistent or pseudo-persistent substances because their entry rate into the environment is higher than the elimination rate". Antibiotics are classified into various kinds, including those that are delivered orally and parenterally have low bioavailability and are eliminated through the gastrointestinal tract, and are utilized in combination therapies for animals. For instance, due to their unabsorbed fraction, tetracyclines can remain in the gastrointestinal tract and expose commensal microbiota for a longer period of time than the treatment period.

6.6 ANTIBIOTIC RESISTANCE IN THE ENVIRONMENT

6.6.1 Aquaculture

It is possible for genetic recombination and exchange to take place in aquaculture, resulting in "genetic reactors" or "hotspots for antibiotic resistance genes", where resistant bacterial populations may emerge. The food sector is said to be among the fastest-growing in the world (Iwu et al., 2020). Because they have antibiotic deposits in their edible tissues, aquatic animals are becoming more and more important (Santos and Ramos, 2016). Among the antibiotics used in aquaculture are erythromycin, florfenicol, sulfonamides, sarafloxacin, and oxytetracycline. When these are introduced into water, the resistant strains of the environmental microbiota are selected for, changing the microbial populations already there (Iwu et al, 2020). Because aquaculture activities discharge heavy metals and biocides into the environment, antibiotic resistance is selected for and spreads as a result.

6.6.2 Manure

Antibiotic residues may be transferred by using animal-based manure on fields for an extended period of time or by directly discharging animal sewage through drainage systems. Antibiotic resistance is greatly encouraged by the use of antibiotics by veterinarians, which is thought to be a major source of antibiotics in the environment (Ezzariai et al., 2018). Excessive levels of antibiotics, between 91 and 136 mg/kg dry matter of tetracyclines, are subsequently released into other environmental niches, such as the soil, surface water, and groundwater (An et al., 2015). Because of its ease of use, effectiveness, and economic viability, aerobic composting is currently one of the most accepted processes utilized in the manufacture of manure (Wang et al., 2016). various microorganism populations, such as algae, bacteria, fungi, protozoa, and protozoa and *Actinobacteria* are present during the composting process, indicating how the process has evolved and performed (Ezzariai et al., 2018). According to Scoton et al. 2016, Composting is the aerobic breakdown by microbes, usually in a controlled environment, of organic materials, such as leftover food, paper, leaves, fruit, and manure.

6.6.3 Agricultural Soil

Antibiotic resistance is highly prevalent in agricultural soil, which has significant effects on public health, food safety, and the economy. After antimicrobials are released into the environment, their fate and behaviour are influenced by soil characteristics such as organic matter content, pH, ionic strength, and capacity to exchange cations, in addition to their physicochemical properties like water solubility, chemical properties, volatility, sorption, and sequestration capacities with climate conditions (Iwu et al., 2020; Albero et al., 2018).

Antibiotics are absorbed into soil by a process called sorption, which is affected by a number of factors such as soil composition, steric configuration, and the antibiotics' amphoteric and amphiphilic qualities. It also regulates the rate at which plants may absorb antibiotics and move about in the soil environment (Albero et al., 2018). Antibiotics that exist in the soil eventually disperse into nano and microparticles. However, their bioavailability and bioaccessibility decline. This leads to sequestration by reducing biological interaction and uptake. The way that plants absorb antibiotics varies depending on the type of plant and the antibiotic. For example, lettuce and carrots have been shown to absorb 2.8–13 μg/kg of trimethoprim and florfenicol, respectively, and onion, corn, and cabbage absorbed 2–17 ng/g of chlortetracycline (Iwu et al., 2020).

6.6.4 Freshwater Bodies

Most antibiotic resistance is located in surface water bodies, including lakes, rivers, and dams. The antibiotics are gathered from many sources, including agricultural, hospital, urban, and rural effluents, and they are combined with native species in the water bodies where these effluents are released.

The following are the two ways that antibiotic resistance spreads in the water bodies (Xu et al., 2016):

1. Multi-drug-resistant strains of bacteria infiltrate the water bodies that are essential to domestic and agricultural activities.
2. Genetic exchange between resident bacteria and contaminating antibiotic-resistant bacteria via phages or integrons takes place in the water, changing the aquatic ecosystem's microbiome.

6.6.5 Irrigation Water

In general, nonpoint pollution sources like agricultural runoffs or close contact with sewage and manure may pollute water used for irrigation. Antibiotic residues are prevalent throughout the environment and have been found in many aquatic systems, including surface water, groundwater, hospital and municipal wastewater, and surface water (Azanu et al., 2016). Groundwater is a hidden resource used for irrigation of crops, which is subject to contamination by human activity and able to endure for a considerable amount of time until the pollution source is removed (Araújo et al., 2017).

6.6.6 Crop Production

Plant-sprayed antimicrobials leak into soil and water systems, affecting people, animals, and the ecosystem. Plant crops, such as carrot, rice, cabbage, potato, corn, spinach, lettuce, tomato, cucumber, and wheat, can absorb and accumulate certain drugs from the growth media via their roots. These medications include amoxicillin, oxytetracycline, trimethoprim, sulfamethoxazole, ofloxacin, sulfamethazine, tetracycline, tylosin, and ciprofloxacin (Iwu et al., 2020). These plant crops are consumed by humans and animals (rodents), which subsequently contributes to the development of antibiotic resistance. Figure 6.3 illustrates the growing risk of antibiotic resistance affecting food crops originating from the complicated agro-ecosystem's nexus.

Antibiotics' biological activity is dependent upon their accessible fraction and their relationship with several environmental factors, including pH, soil type, organic carbon content, water content, and organism type. Figure 6.4 describes the issues of antibiotics and the fate of their metabolites in the environment after excretion. The antibiotics are degraded in the environment by different pathways, which may be biotic and non-biotic. In the biotic method, antibiotics are degraded by microorganisms, and in non-biotic degradation method, the processes include oxidation, reduction, hydrolysis, physicochemical properties, and environmental conditions (Polianciuc et al., 2020). The improper disposal of antibiotics, such as from households or flushing antibiotics down the toilet, could lead to the occurrence of antibiotics in the wastewater system. Patients and individuals need to be specifically educated about the effects of inappropriate drug disposal on the environment as a result of poor pharmaceutical disposal approaches.

6.7 ENVIRONMENT HEALTH

Due to the prevalence of resistant organisms in the environment, it is important to take the "One Health" approach seriously in order to reduce the ongoing development of AMR (Kahn, 2017). This strategy is quite comprehensive and requires a wide viewpoint. Through a risk-based lens, environmental health addresses preventive health and involves the assessment and handling of environmental physical, chemical, and biological components that have the potential to impact human health. As resistant organisms spread throughout the general population and environment, several strategies are employed to prevent and regulate them. AMR is controlled in large part by a number of variables including food safety and hygiene, waste management, and sanitation and hygiene (WASH) (Musoke et al., 2021).

The key environmental health practices for prevention of AMR include:

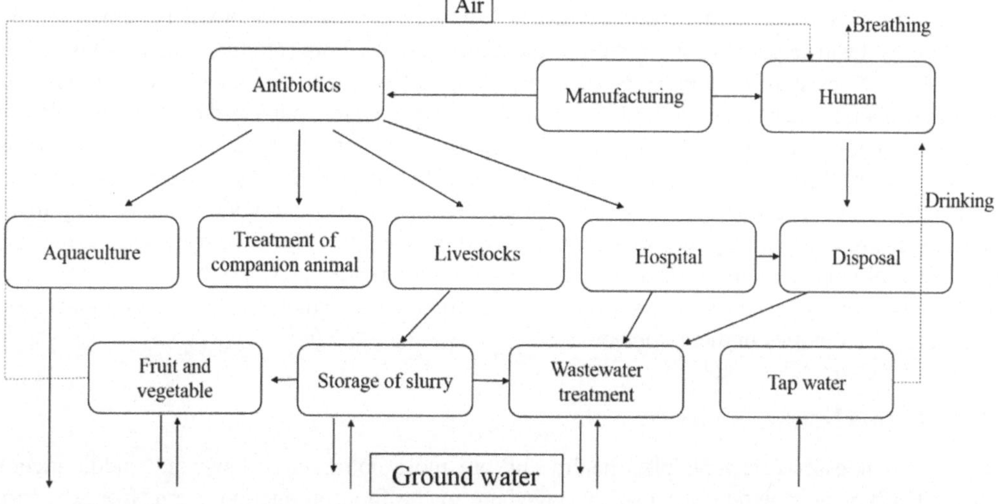

FIGURE 6.3 Antibiotic resistance within and beyond various sections of the agricultural ecosystem.

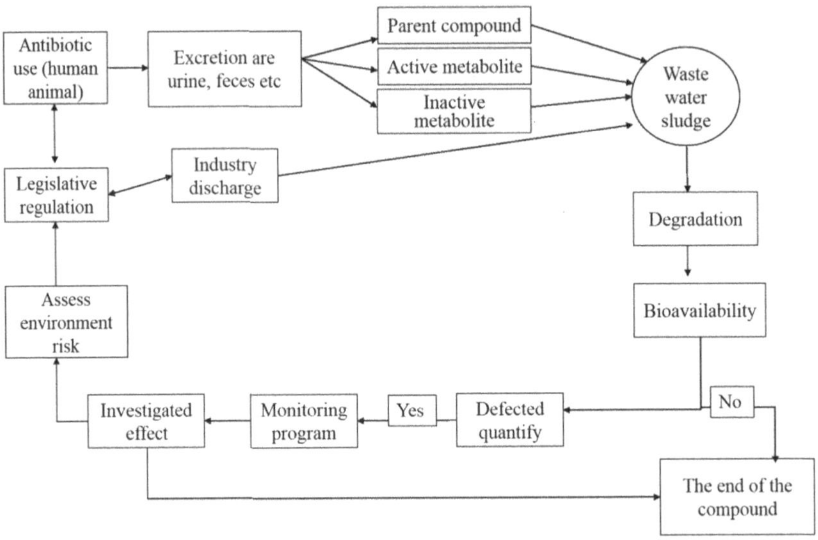

FIGURE 6.4 Antibiotics in the environment from discharge to analysis.

(1) The definition of adequate WASH is having access to and using safe water; disposing of wastewater, including that from medical facilities, and human waste in an appropriate manner; and maintaining good personal hygiene, which includes washing your hands with soap at strategic times to prevent the spread of bacteria that are resistant to antibiotics and the onset of illnesses that may require antibiotic treatment.

(2) Ensuring appropriate food safety and hygiene procedures, such as cultivating vegetables in pure soil and selling and consuming animal products with sufficient antimicrobial withdrawal times.

(3) Appropriate solid waste disposal, which involves getting rid of outdated and unused antibiotics in order to prevent needless exposure to environmental microorganisms.

6.7.1 Water, Sanitation, and Hygiene (WASH)

The majority of WASH-related illnesses, including diarrhoea, cholera, hepatitis A, dysentery, and typhoid, affect low- and middle-income nations. The main risk factor for AMR is human overuse of antibiotics. Insufficient hygienic facilities, including handwashing stations, can contribute to the proliferation of resistant organisms in the environment. Excreta from humans and animals that contain resistant bacteria can readily be washed into surface water sources like rivers and lakes by rain and seep into groundwater, which may result in the spread of infectious diseases that are resistant to treatment (Musoke et al., 2021). Although it is believed that wastewater treatment plants are employed to eradicate a variety of infections, research indicates that antibiotic resistance organisms and residues are not typically removed from them (Manaia et al., 2018). The livestock's faeces and urine eventually release substantial concentrations of antibiotic residues into the environment since they do not normally receive secondary treatment for their waste, which results in the animals' poor absorption of antibiotics in their stomachs.

6.7.2 Food Hygiene and Safety

Food-borne illnesses affect people's health and are more common in low- and middle-income nations. Food contamination can happen anywhere in the food production chain, from the farm to the consumer, and animal feed contributes significantly to the development of resistant germs

(Odeyemi, 2016). Food may contain resistant bacteria if it came from an infected animal or if it was contaminated by insects, filthy hands, or contaminated utensils. Furthermore, the high dosage of antibiotics given to animals raised for food, primarily for growth and development but also for disease prevention, promotes the development of resistant microbes (Musoke et al., 2021). Foods high in antimicrobial resistance, such as milk, beef, and eggs, which are mostly consumed by humans, can increase bacterial resistance to antibiotics (Singer et al., 2016).

6.7.3 Solid Waste Management

Rapid development in low- and middle-income nations has contributed significantly to slum growth and overpopulation, combined with inappropriate commercial business disposal, inadequate home solid waste management, industries, farms, clinics, and hospital settings. Antimicrobials and normal trash are disposed of carelessly, contaminating rivers and lakes among other water sources (Insani et al., 2020). Inappropriate handling of solid waste in rivers, drainage channels, and streets causes flooding and wastewater buildup. This makes it easier for mosquitoes to reproduce and spread diseases including dengue, malaria, and yellow fever. Due to the high rates of morbidity caused by these diseases, antimicrobial medications are used excessively.

6.8 RISK ASSESSMENT FOR HUMAN HEALTH

Evaluating the number of infectious diseases caused by resistant bacteria as a result of environmental antibiotic residues and the number of deaths from such infections is the main objective of the health risk assessment of exposure to antibiotic resistance. To estimate pathogen risks, dose-response assessments – which measure the correlation between pathogen dose and infection probability – are also necessary. The pathogen-specific estimates of morbidity and mortality from antibiotic-resistant infections have been proposed for each pathogenic antibiotic resistance, based on healthcare-associated infections surveys and hospital fatalities (Ben et al., 2019).

6.9 CONCLUSION

Antibiotic resistance as a result of widespread antibiotic exposure is currently a serious public health concern. Antibiotic pollution and the incidence of antibiotic resistance have the potential to directly impact environmental and human health. The most important methods for reducing the prevalence of antibiotic resistance in the agro-ecosystem are improved sanitation and hygiene, limiting the use of antibiotics for purposes other than medicine, and conducting effective biological risk assessments.

REFERENCES

Albero, B., Tadeo, J.L., Escario, M., Miguel, E., and Pérez, R.A. 2018. Persistence and availability of veterinary antibiotics in soil and soil-manure systems. *Science of the Total Environment*, 643, 1562–1570.

An, J., Chen, H., Wei, S., and Gu, J. 2015. Antibiotic contamination in animal manure, soil, and sewage sludge in Shenyang, northeast China. *Environmental Earth Sciences*, 74, 5077–5086.

Araújo, S., AT Silva, I., Tacão, M., Patinha, C., Alves, A., and Henriques, I. 2017. Characterization of antibiotic resistant and pathogenic Escherichia coli in irrigation water and vegetables in household farms. *International Journal of Food Microbiology*, 257, 192–200.

Azanu, D., Mortey, C., Darko, G., Weisser, J. J., Styrishave, B., and Abaidoo, R.C. 2016. Uptake of antibiotics from irrigation water by plants. *Chemosphere*, 157, 107–114.

Bartlett, J.G., Gilbert, D.N., and Spellberg, B., 2013. Seven ways to preserve the miracle of antibiotics. *Clinical Infectious Diseases,* 56(10), 1445–1450.

Ben, Y., Fu, C., Hu, M., Liu, L., Wong, M.W., and Zheng, C. 2019. Human health risk assessment of antibiotic resistance associated with antibiotic residues in the environment: A review. *Environmental Research*, 169, 483–493.

Bondarczuk, K., Markowicz, A. and Piotrowska-Seget, Z. 2015. The urgent need for risk assessment on the antibiotic resistance spread via sewage sludge land application. *Environment International*, 87, 49–55.

Ezzariai, A., Ha, M., Khadra, A., Aemig, Q., El, L., Barret, M., and Pinelli, E. 2018. Human and veterinary antibiotics during composting of sludge or manure: Global perspectives on persistence, degradation, and resistance genes. *Journal of Hazardous Materials*, 359, 465–481.

Founou, L.L., Founou, R.C., and Essack, S.Y. 2016. Antibiotic resistance in the food chain: A developing country-perspective. *Frontiers in Microbiology*, 7, 1881.

Golkar, Z., Bagasra, O., and Pace, D.G. 2014. Bacteriophage therapy: A potential solution for the antibiotic resistance crisis. *Journal of Infection in Developing Countries*, 8, 129–136.

Grenni, P., Ancona, V., and Caracciolo, A.B. 2018. Ecological effects of antibiotics on natural ecosystems: A review. *Microchemical Journal*, 136, 25–39.

Insani, W.N., Qonita, N.A., Jannah, S.S., Nuraliyah, N.M., Supadmi, W., and Gatera, V.A. 2020. Improper disposal practice of unused and expired pharmaceutical products in Indonesian households. *Heliyon*, 6(7), e04551.

Iwu, C.D., Korsten, L., and Okoh, A.I. 2020. The incidence of antibiotic resistance within and beyond the agricultural ecosystem: A concern for public health. *Microbiology Open*, 9(9): 10–35.

Kahn, L.H. 2017. Antimicrobial resistance: A one health perspective. *Transactions of the Royal Society of Tropical Medicine Hygiene,* 111(6), 255–260.

Kraemer, S. A., Ramachandran, A., and Perron, G.G. 2019. Antibiotic pollution in the environment: From microbial ecology to public policy. *Microorganism,* 7, 180.

Kumar, A., and Pal, D. 2018. Antibiotic resistance and wastewater: Correlation, impact and critical human health challenges. *Journal of Environmental Chemical Engineering*, 6, 52–58.

Luyt, C., Bréchot, N., Trouillet, J., and Chastre, J. 2014. Antibiotic stewardship in the intensive care unit. *Critical Care*, 18, 1–12.

Manaia, C.M., Rocha, J., Scaccia, N., Marano, R,. Radu, E. and Biancullo, F. 2018**.** Antibiotic resistance in wastewater treatment plants: tackling the black box. *Environment International*, 115, 312–324.

Mobarki, N, Almerabi, B., and Hattan, A. 2019. Antibiotic resistance crisis. *International Journal of Medicine in Developing Countries*, 40, 561–564.

Murray, C.J., Ikuta, K.S., Sharara, F., Swetschniski, L., Aguilar, G.R., Gray, A., Han, C., Bisignano, P., Rao, P., and Wool, E. 2022. Global burden of bacterial antimicrobial resistance in 2019: A systemic analysis. *Lancet*, 399, 629–655.

Musoke, D., Namata, C., Lubega, G.B., Niyongabo, F., Gonza, J., Chidziwisano, K., Nalinya, S., Nuwematsiko, R., and Morse, T. 2021. The role of environmental health in preventing antimicrobial resistance in low and middle-income countries. *Environmental Health and Preventive Medicine,* 26, 100.

Odeyemi, O.A. 2016. Public health implications of microbial food safety and foodborne diseases in developing countries. *Food & Nutrition Research*, 60, 29819.

Palme, J. B., Kristiansson, E., and Larsson, D.G.J. 2018. Environmental factors influencing the development and spread of antibiotic resistance. *FEMS Microbiology Reviews,* 42, 68–80.

Polianciuc, S.I., Gurzău, A.E., Kiss, B., Ştefan, M.G., and Loghin, F. 2020. Antibiotics in the environment: causes and consequences. *Medicine and Pharmacy Reports,* 93(2), 231–240.

Read, A.F., and Woods, R.J. 2014. Antibiotic resistance management. *Evolution Medical and Public Health*, 1, 47.

Santos, L., and Ramos, F. 2016. Analytical strategies for the detection and quantification of antibiotic residues in aquaculture fishes: A review. *Trends in Food Science and Technology*, 52, 16–30.

Scoton, E.J., Battistelle, R.A.G., Bezerra, B.S., and Akutsu, J. 2016. A sewage sludge co-composting process using respirometric monitoring method in hermetic rotary reactor. *Journal of Cleaner Production*, 121, 169–175.

Sharma, A., Thakur, A., Thakur, N., Kumar, V., Chauhan, A., and Bhardwaj, N. 2023. Changing trend in the antibiotic resistance pattern of *Klebsiella pneumonia* isolated from endotracheal aspirate samples of ICU patients of a tertiary care hospital in north India. *Cureus*, 15, e36317.

Singer, A.C., Shaw, H., Rhodes, V., and Hart, A. 2016. Review of antimicrobial resistance in the environment and its relevance to environmental regulators. *Front Microbiology*, 7(1728).

Viswanathan, V. 2014. Off-label abuse of antibiotics by bacteria. *Gut Microbes*, 5, 3–4.

Wang, R., Zhang, J., Sui, Q., Wan, H., Tong, J., Chen, M., and Wei, D. 2016. Effect of red mud addition on tetracycline and copper resistance genes and microbial community during the full scale swine manure composting. *Bioresource Technology*, 216, 1049–1057.

WHO. 2021. Factsheet: Tuberculosis. https://www.who.int/news-room/fact-sheets/detail/tuberculosis.

Xu, L., Ouyang, W., Qian, Y., Su, C., Su, J., and Chen, H. 2016. Highthroughput profiling of antibiotic resistance genes in drinking water treatment plants and distribution systems. *Environmental Pollution*, 213, 119–126.

Zhang, R., Gu, J., Wang, X., Li, Y., Zhang, K., Yin, Y., and Zhang, X. 2018. Contributions of the microbial community and environmental variables to antibiotic resistance genes during co-composting with swine manure and cotton stalks. *Journal of Hazardous Materials*, 358, 82–91.

7 Infliction of Antibiotic Resistance in the Environment
Challenges and Outlook

Reena Das, Nishant Burnase, Nisha Suryawanshi, Anand Barapatre, and Bikash Kumar

7.1 INTRODUCTION

In the current century, antimicrobial resistance (AMR) is considered to be the most serious global public health threat (Prestinaci et al., 2015). In April 2014, the World Health Organization (WHO) published its first report on global surveillance of AMR using data collected through surveillance networks at national and international levels (WHO, 2014a). Although surveillance systems are not very strong in most countries, the report suggested that AMR exists to a large extent around the globe and is growing at a very high rate (WHO, 2014a).

Resistance to antibiotics has evolved naturally in microbes to overcome the toxic action of antibiotics on their survival. Further, the natural environment plays a key role in the evolution of AMR. Additionally, genetic evolution, mutation, and horizontal gene transfer are key reasons behind the increase in AMR (Davies & Davies, 2010; Hubbard et al., 2019a).

There are several anthropogenic activities, including unregulated consumption of antibiotics, that are possible reasons behind the sudden increase in AMR. Further, in the last few decades, all countries have suddenly experienced a great hike in ovine, bovine, poultry, and fish meat production. About 70% of all antibiotics used for livestock not only engage in fighting against animal infectious diseases but also work as growth stimulators and produce higher yields and higher-quality products. The overdosing of antibiotics for animals (domestic, livestock, and other animals) has significantly to the emergence of multi-drug-resistant strains and difficulties in remedying antibiotic-resistant infections in humans (Talebi Bezmin Abadi et al., 2019). Antibiotics and bacteria resistant to antibiotics leak into local drinking water sources and municipal waste via human waste, which ultimately reaching wastewater treatment plants. Antibiotics used in agriculture also affect the environmental microbiome. It is estimated that about 90% of the antibiotics administered to livestock are excreted through urine and stool into the environment and then usually spread through fertilizer, groundwater, and surface runoff (Bebell & Muiru, 2014; Lee Ventola, 2015).

The AMR is being developed by a combination of public behaviour, technological advancements, surrounding environment, and genetic as well as structural changes in microbes. These changes comprise a positive difference in the number of vulnerable hosts, international travel and markets, alterations in technology as well as in industries, microbial adaptation power and genetic change, and, most influencing the collapse of public health measures (Cohen, 1997). Antibiotics are recommended for the patient based on his condition, and their negative effects extend far beyond the patient themselves. Under the influence of economic development and a drastic increase in the biomass production of animal protein for human consumption, antibiotics are the most common contaminants of locally produced and imported meat and poultry (Bebell & Muiru, 2014). Historically, human migration and globalization have strongly contributed to the development and

DOI: 10.1201/9781003268925-7

spread of infectious diseases. The United Nations World Tourism Organization (UNWTO) reported in 2018 that approximately that 19% of the world's population moved internationally, with about 11 million people travelling to medical facilities to pursue reasonable healthcare. Immigration and international travel of effectively moved people can contribute to the global transmission of AMR (Bokhary et al., 2021).

Surveillance is one of the most important aspects of preventing future AMR-associated pandemics. Global participation in initiatives such as the Global Antimicrobial Resistance and Use Surveillance System (GLASS) can be crucial in achieving the objectives of prevention through surveillance (Johnson, 2015; Opintan et al., 2015; Pfizer, 2020; WHO, 2014a). Data analytics and artificial intelligence can help in combating high AMR and devising policies for future implementation of AMR preventative measures (Rabaan et al., 2022). This chapter will address the pervasiveness of antibiotic resistance in soil and water environments, its evolution, transmission, and understanding of the resistance crisis. Potent steps must be taken in terms of the spread of resistance genes in environmental bacteria into human pathogens, to decrease the spread of resistant bacteria to people and via wastes and water, to minimize the levels of antibiotic consumption, and to improve the management of waste containing antibiotic residues. This chapter will also provide a perspective on methods of surveillance of the factors contributing to resistance and identifying elements of artificial intelligence to alleviate the transmission and resistance of antimicrobials.

7.2 EVOLUTION OF AMR

7.2.1 Role of the Natural Environment in AMR

Natural environments play an important role in the evolution of antibiotic resistance in the microbiome. Antibiotic resistance is known to have emanated in the environment particularly due to two factors, which are mutations in the pre-existing genome and horizontal gene transfer (HGT). Since pathogenic microbes were earlier reported to be sensitive to antibiotics, before the decades of incessant usage of antibiotics, the concept of resistance acquired via HGT lies in the annex of genes in the environmental communities of different bacterial populations or species (Martinez, 2009). The feature that makes the environmental microbiome more prone to the transfer of resistance factors compared to human and animal microbiota is the gigantic reservoir of the accessible gene pool (Larsson & Flach, 2022). Also, antibiotics are commonly produced by environmental bacteria, and by some means, they do not let the antibiotic affect them. This enforces the evidence of the unfolding of HGT-acquired antibiotic resistance via the environment (Shintani et al., 2020).

Antibiotic-resistant genes (ARGs) evolution is mostly acquired because of recent antibiotic usage for human therapy or farming. The stepwise evolution documents the ability of ARGs to move within the genome with the help of insertion sequences, transposons, gene cassettes, and other factors such as plasmids and integrative conjugative elements (Jian et al., 2021). The genetic elements typically involved in the transmission of ARGs are abundant in soil and water environments polluted with faecal matter and other human-based anthropogenic activities and contribute to the spread of resistance genes. Apart from this, there are many instances of common characteristics in human commensal bacteria and some opportunistic environmental bacteria proving that resistant elements are archaic and acquired naturally; for example, the presence of chromosomally encoded multi-drug-resistant (MDR) efflux pumps and ampC β-lactamase gene in all strains of *Pseudomonas aeruginosa*, which provides intrinsic resistance by displaying low susceptibility to antibiotics (Hubbard et al., 2019b).

Understanding the role of each mechanism involved in imparting resistance to antibiotics has been further discussed in the next few sections.

7.2.2 Mechanism of Resistance

7.2.2.1 Mutation

To attain resistance against antibiotics, environmental microbiota involves responses either by using its pre-existing resistance machinery or by mutation in pre-existing genes or acquired genes through HTS.

Generally, pre-existing resistance has been reported in organisms such as *Streptomyces* sp. recovered from soil and usually employs mechanisms such as inactivating resistance mechanisms, efflux execution, and target modifications (Davies & Davies, 2010). However, many new insights into antibiotic resistance evolution in natural microbiota have been noted in the recent past. For instance, the β-lactamase gene is recognized for its universal distribution and random mutations associated with it. Novel types which have been unearthed recently, such as the new β-lactamase CTX-M acquired in *Kluyvera* strains, are known for their hydrolyzing activities of cephalosporin. This highly transmissible gene has an efficient and rapid mode of HGT and is a global threat. Other acquired resistances, such as macrolide related to rRNA modifications and the fluoroquinolone resistant *qnrA* gene related to the modification of a secondary amine by aminoglycoside N-acetyltransferases, impart a low-level tolerance favouring resistance mutations in genes otherwise lacking in intrinsic resistance (Meng et al., 2022).

7.2.2.2 Intrinsic Transfer

Intrinsic resistance is usually developed by the presence of inherent structural or functional properties towards certain antibiotics. The presence of an antibiotic-resistant phenotype in the bacterial genome, which can be effectively studied using bacterial genome sequencing, has given a clear picture of many potential/intrinsic gene functions in bacteria that may lead to resistance phenotypes in different environments. The creation of partial mutant libraries of the bacterial genome using saturation mutagenesis aids in identifying mutants sensitive to antibiotics, and overexpressing these wild-type genes constitutes a resistance phenotype (Figure 7.1). This approach has been universally endorsed for the identification of potentially resistant genes (Canteón, 2009).

An exclusive collection of all antibiotic-resistant genes called an "antibiotic resistome" was conceptualized for the microbiota of a particular environment. The resistome includes precursor genes as well as modified genes that evolved over a particular period as resistance elements (Wright, 2007). The study of resistome includes constructing a library of AMR microorganism strains to investigate antibiotic resistance, screening them against different antibiotics, and identifying the

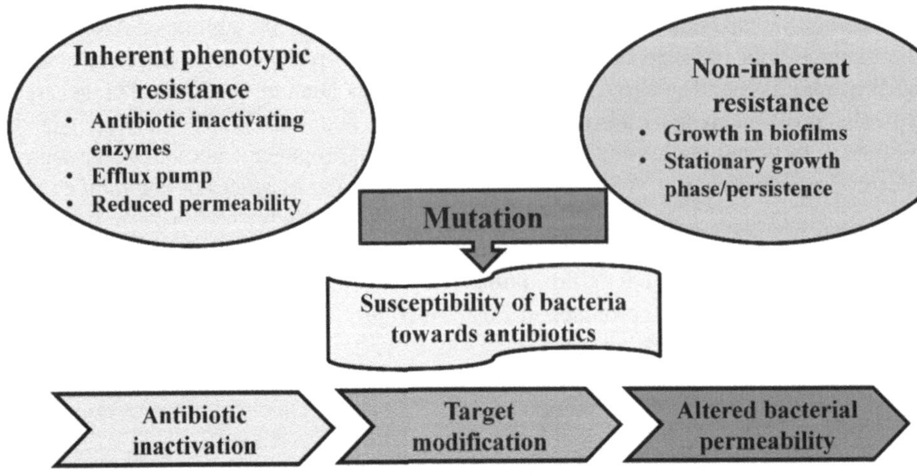

FIGURE 7.1 Rudiments of acquired resistance of bacteria towards antibiotics.

genes administering resistance mechanisms. Usually, the already existing genes produce antimicrobials. This analysis helps to identify not just the resistance mechanisms affecting natural, but also synthetic and semi-synthetic, antimicrobials. This mechanism is not just profound in studying one organism but also surrounding organisms in nearby proximities (metagenomic) to understand resistance gene transfer and is termed an "antibiotic extended resistome" (D'Costa et al., 2007).

7.2.2.3 Other Mechanisms such as Altering the Functions of Antibiotics

Many bacteria are known to nullify antibiotics such as β-lactams, cephalosporins, aminoglycosides, phenols, and macrolides by hydrolyzing them with the help of enzymes. Bacteria produce enzymes that can add groups such as acyl, phosphates, and nucleotides to the susceptible sites of antibiotics, preventing them from binding to the target proteins. The enzymes acetyltransferases, phosphotransferases, and nucleotidyltransferases are known to work on the exposed hydroxyl and amide groups of aminoglycosides and render them susceptible to modifications, leading to resistance toward modified antibiotics (Nordmann et al., 2011).

7.3 ANTIBIOTIC-RESISTANT GENES AND THEIR PERVASIVENESS

Indigenous bacterial strains of phyla such as *Proteobacteria*, *Actinobacteria*, *Firmicutes*, *Bacteroidetes*, etc. in various environments have been found to have resistance to antibiotics ampicillin, chloramphenicol, erythromycin, streptomycin, and tetracycline. The multifariousness of ARGs in different environments is either due to the natural diversity or heavy human intervention (Zhang et al., 2021). In the recent past, workers have developed many drugs such as polymyxin for curbing infections by MDR Gram-negative bacteria and colistin against the *Enterobacteriaceae* family, including *Klebsiella* species, *Escherichia coli*, *Shigella* species, *Enterobacter* species, and *Salmonella* species. The major route of obtaining resistant ARGs in the systems is through HGT. Processes such as transduction, splicing, and transformation constitute the footing of HGT. Many drug-resistant bacteria found naturally in the soil act as hosts for HGTs for multi-drug resistance genes (MRGs), which include organisms such as *Xanthobacteraceae*, *Bacillus*, *Actinobacterium*, and *Geobacter*. Strains such as *Pseudomonas*, *Acinetobacter*, and other Gram-negative bacteria have been actively evolving as MDR as they have been reported to possess resistance against spectrum β-lactamases, carbapenemases, and mucilage resistance (Meng et al., 2022).

7.3.1 Transmission of Resistance Genes

Environment plays an integral part in the event of genetic transmissions across bacterial species, by being the repository for the ARGs and contributing to the large-scale spread of resistant genes. Mobile genetic elements (MGEs) grant the movement of genes with the help of insertion sequences (IS), transposons (Tn), and gene cassettes as well as plasmids and integrative conjugative elements. These accessory genetic elements expedite horizontal gene transfer and acquisition of ARGs. Apart from intracellular conjugation through genetic elements, natural transformation through bacteriophages, splicing, and transduction are common factors for bacterial HGT (Meng et al., 2022). Vertical Gene Transfer (VGT) also offers various modes of proliferation and spread of ARGs. A wide variety of drug-resistant genes have been identified as vectors of the spread of ARGs, including *tet*, *qnr*, *sul*, and *intI*.

7.3.1.1 Intracellular Conjugation Mediated by Plasmid through Splicing

Conjugation via transferring plasmids through pili on the cell membrane is considered the most dominant mechanism of transmission of ARGS (Li et al., 2019). Plasmid-mediated transmission of resistant genes and other elements is by far the most prevalent mode of HGT. Apart from the resistant gene, the plasmids carry mobile genetic elements such as IS and Tn. Plasmids are transferred through splicing of the ARGs to the recipient cells belonging to either identical or unrelated bacterial species. The ARGs are aggregated and expressed as Type IV secretion systems or T4SS during the splicing of the donor strains. Apart from the genes, the membrane-bound Type IV coupling proteins (T4CPs), which are DNA-dependent ATPases and relaxosome proteins such as TraI, TraM, and TraY, contribute towards nicking and translocation and thus translocating the genes. However, the splice transfer process within the receptor cells requires relaxosomes from the acceptors and T4CPs from the donors for the facilitation of translocation (He et al., 2019). Plasmids in general have diverse genetic characteristics portraying wide varieties of molecular responses and regulatory systems to undergo gene transfer.

7.3.1.2 Natural Transformation Mediated by Bacteriophages

Bacteriophages carrying resistant genes constitute the second most common phenomenon of HGT. However the concurrent insertion sequence (IS) and transposons (Tns) remain intact with bacteriophage, making them a good mobile-resistant element. Phage-mediated transduction is either specific/specialized or universal/generalized. Specific transduction involves the uptake of a distinct gene segment, whereas universal defies any distinctiveness, and any segment of the bacterial genome can be moved, mostly displayed by pac-type phages. A rather newer mode of transduction called lateral transduction has been reported for the displacement of larger segments of the genome with higher frequencies, deciphering the rapid increase in gene transfer in the recent past (Chen et al., 2018). In generalized transduction, usually after induction, the phage either undergoes lytic or lysogenic conversion. During the process, the phage capsid containing the phage DNA is released to the recipient cell, thus facilitating uncomplicated ARG acquisition from neighbouring cells. The survival of the ARGs in the recipient cells is determined by the intensity of integration into the bacterial genome. In specific transduction, the integration frequency is higher owing to the integrase gene, whereas in universal transduction recombination of the exogenous gene into the host chromosome determines the efficiency of integration of ARGs (Brigulla & Wackernagel, 2010). MDR characteristics in *S. aureus* (Haaber et al., 2016) and *Salmonella's* fluoroquinolone exposure are approved examples of it (Bearson & Brunelle, 2015).

7.4 ENVIRONMENTAL TRANSMISSION

The prominent drivers of ARG's emergence include clinical, agricultural, and natural antibiotics, accompanied by biocides, heavy metals, and fertilizers. These compounds are predominantly transmitted into the environment via municipal and industrial wastewater, livestock manure, sewage sludge, agricultural lands, and aquaculture setups (Samreen et al., 2021). Traditionally, human infectious diseases have been treated with antibiotics; however, excessive use of antibiotics can lead to gut dysbiosis and the selection of resistant strains (Ahmed & Gareib, 2016; Sohail et al., 2016). Hospital effluents are a significant site for horizontal gene transfer, which favours the transfer of antibiotic resistance among pathogens. Other factors contributing to the spread of ARGs among bacterial species are mobile genetic materials, such as transposons (Bruno & Carolissen Mackay, 2012; Ragan & Beiko, 2009). Antibiotic residues have been traced in chicken, meat, and milk because of the use of veterinary medicines as prophylactic agents and feed additives. Further, these antibiotics are released into animal manure, which is then passed on to agricultural lands and, ultimately, to people through the food chain (Beyene, 2015; Gandra et al., 2015). The mechanism of antimicrobial resistance shares structural and functional similarities with that of heavy metals when they act as drivers of AMR. Arsenic, zinc, cobalt, manganese, and silver are common examples of heavy metals that cause AMR (Hobman & Crossman, 2015; Pal et al., 2017). Biocides are microbiocidal substances frequently found in cosmetics, medical supplies, household cleaners, and wipes. A high concentration of biocides promotes the selection of mutations that lead to antibiotic resistance (Bengtsson-Palme & Larsson, 2016; Maillard, 2018).

ARGs develop antimicrobial resistance in human pathogens transmitted from the environment to clinics via selective flow. The AMR drivers contaminate agricultural lands and water bodies and enter the food chain via crops, fish, and animal food sources (Figure 7.2). Consumption of contaminated food leads to the emergence of antimicrobial resistance or the transfer of ARGs to the natural gut microflora. Finally, AMR is transmitted to human pathogens, resulting in antimicrobial-resistant infections (Skandalis et al., 2021).

7.5 RESISTANCE IN THE ENVIRONMENT

Antibiotics are used in different fields of life sciences, *viz.* medicine as a disinfectant, veterinary medicine in animal husbandry as a growth promoter, fish farming, and some other areas, and are at last released into the environment after their use. In the last few decades, numerous pathogenic bacteria present in our surroundings have evolved significant resistance to antibiotics, and their ecotoxic effects have become one of the primary concerns for humans and ecology. Antimicrobial resistance is a significant public challenge in all three dimensions (air, water, and land), contributing to its evolution and spread (Kümmerer, 2004; Liguori et al., 2022). The antimicrobial traces present in the surrounding environment represent an increasing threat because of their high biological activity and accumulating power by which they can produce different antagonistic effects in target as well as in non-target species. Antimicrobials are the third-largest cluster among all pharmaceuticals in human medicine and the most noticeable category in veterinary medicine, with the expected annual worldwide consumption between 100,000 and 200,000 tons (Senta et al., 2013).

Metabolically, human beings and animals rapidly digest antibiotics, and it is estimated that around 90% of some antibiotics are generally excreted as the unmodified, completely active parent molecule as well as partially metabolized forms through faeces and urine. The primary sources

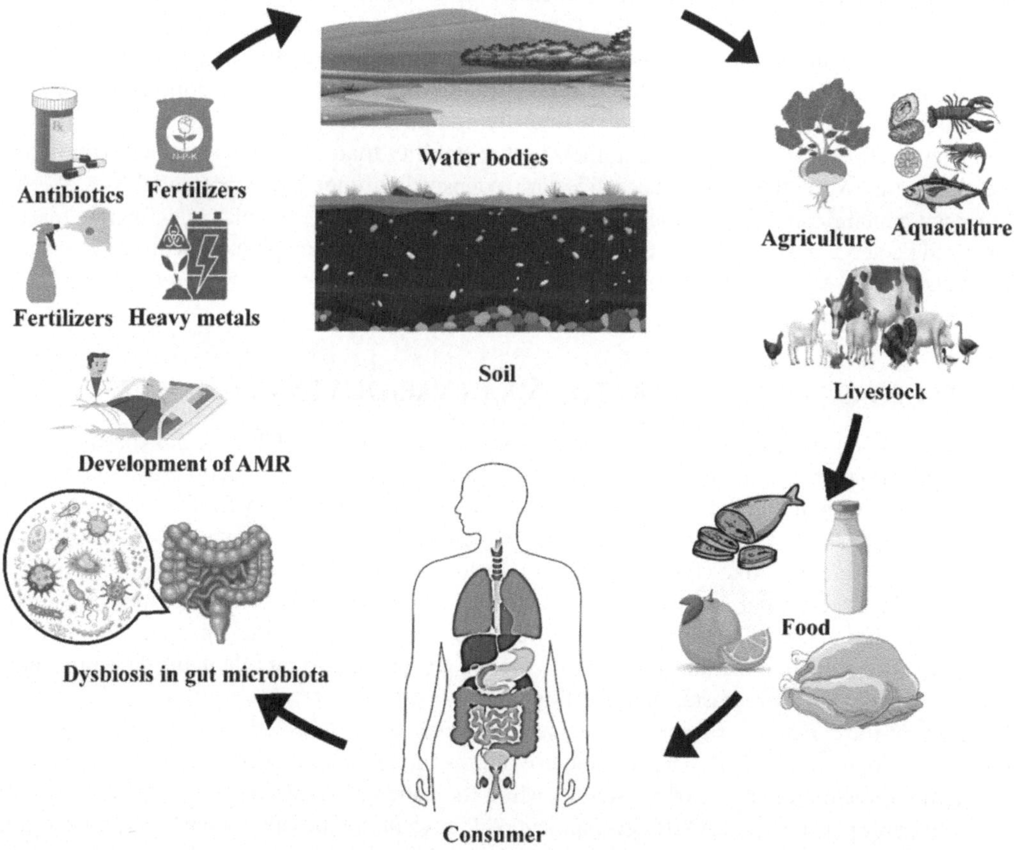

FIGURE 7.2 A selective flow of AMR emergence and transmission.

of these entire active molecules and partially modified metabolites of antibiotics are the wastewater generated by hospitals, outpatient clinics, and veterinary clinics. These municipal wastewater treatment plants might be a possible source of antibiotic contamination in the soil and aquatic environment (K. Kumar et al., 2005; Szymańska et al., 2019; Taneja & Sharma, 2019). Antibiotic resistance in the environment is considerably high in hospital effluents and sewage treatment plants, although hospital effluents only account for less than 1% of the total amount of municipal sewage. In addition, the amount of antibiotics found in municipal sewage and sewage treatment plants (STPs) is 100 times lower compared to hospital effluent (Kümmerer, 2009b). The manure produced by cows, swine, chickens, and turkeys is another source of antibiotic contamination in the soil and water. Antimicrobial agents are fixed in the ground by association with soil chemicals and manure. The most common antibiotics in this manure are tetracyclines (oxytetracycline and chlortetracycline), tylosin, sulfamethazine, amprolium, monensin, virginiamycin, penicillin, and nicarbazine (K. Kumar et al., 2005).

In the agricultural catchment areas, water is contaminated mainly by soil containing antibiotics, which originate from slurry and sludge used as fertilizers. Compounds released into the soil system can subsequently be leached into surface water or groundwater, cycled, and recycled within the environment. Overspills from agricultural land or seepage from septic tanks and broken sewage pipes are notable sources contributing antibiotic-resistant bacteria into groundwater (Kümmerer, 2009b; Wellington et al., 2013). The drug-manufacturing units are also major polluters of antimicrobial compounds in the environment through their plant waste effluents. The available technologies used in the treatments and treatment units for recycling the generated effluents are not very efficient at neutralizing these generated pollutants. There are enough antibiotics, resistant microbes, and resistance genes left in the environment to release effluent (Tamhankar & Lundborg, 2019).

Fish farming industries are also one of the sources of drug-resistant bacteria, where antibiotics are applied for therapeutic purposes and as a feed additive. Antibiotics that are mainly used in the fish industries are especially sulfonamides (sulfamethoxazole and trimethoprim), macrolides (erythromycin and clarithromycin), as well as fluoroquinolones (ciprofloxacin and norfloxacin) (Kümmerer, 2009a). In contrast, due to their higher stability than other classes of antibiotics, sulfonamides, fluoroquinolones, and macrolides are extensively grouped as important classes of environmental contaminants (Senta et al., 2013). Antimicrobial resistance has also been seen in marine colonized bacteria and the bacteria found in estuaries or coastal waters polluted with sewage water. It was also seen that *E. coli* isolates carrying antimicrobial drug resistance were found in bird residents at isolated places in the Arctic Sea (Kümmerer, 2004, 2009b).

7.6 FACTORS CONTRIBUTING TO AMR IN VARIOUS ENVIRONMENTS

The Organisation for Economic Co-operation and Development (OECD) reports that in human healthcare, about 50% of antibiotics from whole antimicrobials are used incorrectly, and in long-term care facilities and ambulatory care, this use might be as high as 90% (D'Atri et al., 2019). The emergence of antimicrobial resistance depends on many factors. According to the Institute of Medicine report, factors like changes in human demographics and behaviour, changes in technology and industry, economic development, international tourism and business, change and adaptation in microbes and the failure of public health procedures influence the emergence of AMR in microbes. Different combinations of the above factors play an important role in the development of AMR in specific microorganisms (Cohen, 1997).

As one of the end users, the community has also played a crucial role in antibiotic consumption and the development of AMR. The sensible use of prescribed antibiotics plays a vital role in minimizing the growth and spread of resistance, while the patients' knowledge and involvement with antibiotics also play a part in AMR development. Although antibiotics are recommended according to the individual patient's condition, unlike other drugs, antibiotics can show their effects far beyond the individual. Therefore, controlled consumption of antibiotics requires the involvement of experts and committed healthcare practitioners as well as the public. Strict monitoring and supervision will also be required while using antibiotics and their related compounds. Consequently, it is essential to define at what level the community understands antibiotics and their proper use. Discourage self-medication, and education on hygiene and disease transmission, also has a great impact on stopping the emergence and spread of AMR (Bebell & Muiru, 2014; Di et al., 2022; Okeke et al., 2005).

In the last decades, at the global industrial level, individual earnings of low- and middle-income countries have rapidly increased resulting in the demand for animal protein for human consumption also increasing. As a result, there has been a drastic increase in the biomass production of animals. To fulfil this requirement, these low- and middle-income countries have adopted highly cost-efficient and integrated livestock production systems, in which antimicrobials are required to keep the animals healthy and maintain productivity. Increased antimicrobial consumption ultimately results in the rise of antimicrobial resistance (van Boeckel et al., 2015). These low- and middle-income countries, also having weak health systems, are particularly vulnerable to an increase in AMR. In these countries, mortality has increased by more than 90% due to AMR, where these countries are already fighting several pre-existing problems such as infectious diseases, poor sanitary standards, limited availability of active antimicrobials, and poor health infrastructures. Evolving resistance to cures for other diseases, such as tuberculosis, malaria, and HIV, has huge impacts in lower-income countries. The rising incidence of drug-resistant strains of tuberculosis is well known. It is estimated that around 480,000 new cases were seen in 2013, most of which remain untreated. The same situation of spreading resistant strains of malaria is also observed, and the development of resistance to antiretroviral therapy for HIV is strictly examined. Most of the healthcare facilities in these countries, especially in sub-Saharan Africa and some parts of Asia, are not able to perform simple basic diagnostic testing to reveal the possible reasons for a disease where only antibiotic susceptibility tests are required. This weak diagnostic infrastructure has an impact not only on clinical care and hospitality decision-making but also extends to failure to apply suitable surveillance of AMR at regional or national levels (Nkengasong et al., 2010; O'Neill & Team, 2014a; Semret & Haraoui, 2019; WHO, 2014b).

International travel has also been recognized as a critical contributor to the transmission of AMR, AMR genes, and AMR-associated organisms globally. In most cases, high-income countries that receive the AMR are more likely to have originated from middle- and low-income countries. Some recent studies have shown that almost 30% of tested travellers were infected with an acquired AMR bacterium during international travel. Asia is the most common origin of resistant bacteria in travellers, accounting for about 36% of the total isolates. In addition, among the overall drug resistance, β-lactams ($\approx$ 35%) and quinolones ($\approx$ 31%) were the most recognized drug-resistant organisms. Epidemiological studies found that there are three major risk factors: travel destination, antimicrobial usage, and travellers' diarrhoea. Regarding general travel, medical tourism was two-fold as likely to be associated with multi-drug-resistant organisms (Bokhary et al., 2021; Sridhar et al., 2021).

The environment will also play an essential role in the development of AMR. It facilitates the transmission of ARGs (antimicrobial resistance-associated genes) between the environment, human, and animal-associated bacteria. It also acts as a pool or transitional habitat between the resistant bacteria and ARGs. Moreover, it plays a critical part in the progression of novel resistance factors by selecting resistance in combination with genetic diversity by which bacteria can develop ARGs (Bengtsson-Palme et al., 2021). One more factor, which helps in microbial adaptation to generate resistance, is the increased use of broad-spectrum antimicrobial agents. Nowadays, these agents are frequently used in clinical practice as a substitute for accurate diagnosis or to increase the chance of therapeutic success. Moreover, this frequent use increases the scope of selective pressure on the microbes to develop AMR (Cohen, 1997).

7.7 EFFECT OF AMR/ENVIRONMENTAL DAMAGE MEDIATED BY AMR

AMR poses a serious threat to people, animals, plants, and the environment on a worldwide scale. Comparatively less attention has been paid to the environmental impact of antimicrobial resistance than to AMR in animal and human health. Antimicrobial resistance is, nonetheless, widely distributed in the natural environment. People, animals, food, and the environment (in water, soil, and air) all contain drug-resistant microorganisms. AMR development and spread may be most prevalent in water and also in the soil, especially in areas with poor water supply, sanitation, and hygiene (Medlicott et al., 2020) "Superbugs" can transfer from person to person and from animal to animal, as well as from animal products to the human diet. There is compelling evidence that the environmental discharge of antimicrobial substances, together with close contact between naturally occurring bacterial populations and released resistant bacteria, is causing bacterial evolution and the rise of more resistant strains (UNEP, 2017).

Antimicrobial usage has complicated environmental effects. Residues of antimicrobial medications, as well as resistant organisms, may be present in drinking and recreational water. Even without ever receiving antibiotic treatment, wildlife that comes into contact with output from wastewater treatment facilities or livestock farms where antimicrobials are utilized might become colonized with drug-resistant microbes (UNEP, 2022).

7.7.1 Antibiotics Consumption and the Release of Antibiotics into the Environment

It has been noted over the decades since the discovery of antibiotics that the use of antibiotics in human medicine, veterinary medicine, and agriculture is connected to the contamination of various environmental compartments (such as surface water, groundwater, drinking water, municipal sewage, soil, vegetables, and sludge), which in turn leads to a rise in antibiotic resistance and has detrimental ecological effects (Hanna et al., 2018). Additionally, the use of antibiotics promotes the growth of bacteria or genes that are resistant to them, increasing the risk of transmission to humans from the environment. It is believed that greater usage of antibiotics may result in infections with bacteria resistant to antibiotic therapy, lengthening sickness, and increased morbidity and mortality in humans (FAO et al., 2018; Merlin, 2020).

According to recent studies, 33,000 people in the EU pass away each year from infections with bacteria resistant to last-resort antibiotics like carbapenems and colistin, accounting for 39% of all antibiotic-related deaths (ECDC, 2018). Similar circumstances exist in the United States, where 2.8 million people contract an antibiotic-resistant infection each year, and more than 35,000 of them pass away (CDC, 2022) Consequently, antibiotic-resistant illnesses are extremely expensive for the economy and healthcare system. Due to the need for last-line antibiotic therapy, which is significantly more expensive than first- and second-line antibiotic therapy, the cost of treating these infections has increased. Additionally, a patient with illnesses that are resistant to antibiotics may need to stay in the hospital for an additional 6.4 to 12.7 days. According to estimates, infections with antibiotic-resistant bacteria cost the EU and the United States a combined total of $55 billion annually in lost productivity and healthcare costs (Mobarki et al., 2019).

7.7.2 India's Challenges with AMR

India is now being recognized as "the world's AMR capital" (Chaudhry & Tomar, 2017). While the introduction of more advanced, multi-drug-resistant (MDR) pathogens presents fresh hurdles for diagnosis and treatment, India is still working to fight off the oldest enemies like cholera, malaria, and bacteria that cause tuberculosis, which are getting even more drug-resistant. The issue is further complicated by elements like poverty, illiteracy, overpopulation, and starvation (Swaminathan et al., 2017). The general public's ignorance about infectious diseases and lack of access to healthcare frequently prevent people from seeking medical guidance (Laxminarayan et al., 2016). This frequently results in the self-prescription of antimicrobial drugs without any awareness of the dosage or length of the recommended course of action (Morgan et al., 2011a). Due to a lack of adequate diagnostic tools for determining the pathogen and its drug susceptibility, many people who seek medical guidance wind up receiving broad-spectrum, expensive antibiotics. The proliferation of MDR microbes in healthcare centres is aided by low ratios of patients to nurses and doctors as well as improper infection control and prevention measures protocol (Laxminarayan & Chaudhury, 2016; Swaminathan et al., 2017). Easier access to over-the-counter (OTC) medications has an impact on AMR (Bate et al., 2009).

Antimicrobial Resistance causes more deaths than cancer, traffic accidents, and other causes combined, with 700,000 individuals losing the battle each year and an additional 10 million anticipated to do so by 2050 (O'Neill & Team, 2016). According to economic estimations, AMR would cost the world \$100 trillion by 2050 due to a 2–3% GDP decline and a 3–8% decline in livestock (O'Neill & Team, 2014b). Various parties, including the WHO, have been concerned about the global spread of AMR. AMR was declared an immediate top priority by the WHO (WHO-ROSEA, 2011), and various international leaders from Europe have described their guidelines for the prevention of AMR (DOH-UK et al., 2013; Smith et al., 2016). Genuine attempts are being made to combat this threat. In April 2017, India also launched a Nationwide Implementation Plan for AMR, describing the suggested measures over the upcoming five years (MoHFW, 2017b).

It is critical to understand that AMR is a complex issue that necessitates a multifaceted strategy for control. The One Health assumption highlights the linkages between people's well-being, animal care, food safety, and environmental health, and encourages cooperation between the health agencies responsible for these areas (Dahal et al., 2017). It is essential to give each of these contributors the attention they deserve to effectively regulate AMR. However, the majority of investigations have focused on the usage of antibiotics without regard to the effects they may have on human and animal health (O'Neill & Team, 2016; WHO-ROSEA, 2011, 2017). The majority of AMR caused by the environment has thus far gone unrecognized. The situation is further complicated by the fact that one of the factors contributing to AMR is the environment, which varies greatly between different geographic regions of the world. As a result, any developed country's estimates (such as England) and suggestions might not even be appropriate for the rest of the world (Fletcher, 2015; Taneja & Sharma, 2019).

The advantages and necessity of using antibiotics cannot be disputed, but it is also important to consider how antibiotics affect the environment because they can spread resistance and exhibit several unfavourable impacts on the ecosystem as a whole. To better understand how antibiotic residues affect the environment, preventative actions must be taken as soon as possible to prevent later detrimental consequences.

7.8 ANTIMICROBIAL SURVEILLANCE

Antimicrobial surveillance refers to actions taken to track shifts in microbe populations to better comprehend the development of anti-infective resistance patterns. It may be carried out on a national, international, local, or medical training centre basis (Pfizer, 2020).

7.8.1 Why Is Surveillance Crucial?

Surveillance is crucial in all areas for the prevention of infections and the management of microbial resistance in relation to health since it offers the data needed to create and track therapeutic standards, the use of innovative antimicrobials and vaccines, antimicrobial stewardship programmes such as "Be Antibiotics Aware", community health initiatives, and infection prevention regulations.

Pathogens adapt and find ways to counteract the effects of anti-infectives, which is what is referred to as antimicrobial resistance. It is commonly acknowledged that AMR, particularly in

Gram-negative bacteria, is one of the major risks to global public health today, resulting in 700,000 fatalities per year (Pfizer, 2020).

Clinicians and public health officials must use surveillance as a key tool to stop the spread of AMR by delivering early notice of impending dangers and assisting decision-makers in taking action before they worsen, and by coordinating plans for infection prevention and control and public health policy (Uchil et al., 2014).

Effective chemotherapeutic medicines must be administered promptly for dangerous infections to be successfully treated (Lange et al., 2014). While some infections, like tuberculosis (TB), whooping cough, and gonorrhoea, are brought on by a single pathogen, most infections, including those that affect the skin and soft tissues, the upper and lower respiratory tracts, the urinary tract, meningitis, and sepsis, are brought on by a variety of pathogens. Thus, to make clinical decisions on initial treatment, it is necessary to be aware of the most likely pathogen(s) and their propensity for antibiotic susceptibility (Bloom et al., 2017). While practical experience over time plays a role in the acquisition of some of this knowledge, surveillance offers a more rigorous and objective method. Surveillance, according to the US Centers for Disease Control and Prevention (CDC), is "the ongoing systematic collection, analysis, and interpretation of health data essential to the planning, implementation, and evaluation of public health practice, closely integrated with the timely dissemination of these data to those who need to know". Ibrahim (2020) suggested, surveillance is the process of gathering information and promptly disseminating it to strategic planning and action. These definitions imply that conducting surveillance calls for an easily accessible data source. Microbiology labs regularly identify and assess the susceptibility or resistance of bacteria obtained from clinical specimens, and this information is used for the surveillance of antibiotic resistance. These findings are kept in the computer system of the lab and, if accessible, gathered, and analyzed, can provide information about the level of antibiotic resistance observed in various bacterial species or isolates from various types of infections (Johnson, 2015). Antibiotic resistance can be tracked for changes or variations over time and across geographic boundaries.

7.8.2 Obstacles to Efficient AMR Surveillance

7.8.2.1 Inadequate Lab Infrastructure

The fundamental needs for a functioning laboratory infrastructure are not met in LMICs (Bebell & Muiru, 2014; Fitzgibbon & Wallis, 2014; Ombelet et al., 2018). These include problems with the water supply, electricity, light sources, climate control, and ventilation, as well as biosafety regulations, slow internet connectivity, and coverage, a lack of soap and alcohol gel, dust, inadequate restroom facilities, and poor construction that makes deep cleaning difficult (Ombelet et al., 2018). Pathogen detection and antimicrobial susceptibility testing can be less accurate and of lower quality and dependability in laboratories with inadequate infrastructure (Figure 7.3) (Dar et al., 2016; Opintan et al., 2015).

7.8.2.2 Limited Staffing and Training Resources

The effectiveness of data management is impacted by understaffing, a lack of devoted employees, and a shortage of clinical and laboratory personnel who are trained (Ashley et al., 2019; Boerma & Mathers, 2015; Jayatilleke, 2020; Ombelet et al., 2018). There are few specialists in the field among microbiologists and healthcare workers (de Kraker et al., 2016). Another significant issue is the absence of recognized professional standards or a profile for clinical microbiologists and postgraduates (Ombelet et al., 2018). The limited participation of microbiologists in staff orientation and training, the absence of governance and leadership, problems with after-hours calls, and inadequate management are difficult impediments to overcome (Figure 7.3). The focus on activity, advocacy, and championing is provided by leaders (de Kraker et al., 2016). Lack of direction and logistical management for stock, waste, and inventory control affects the effectiveness of the activity.

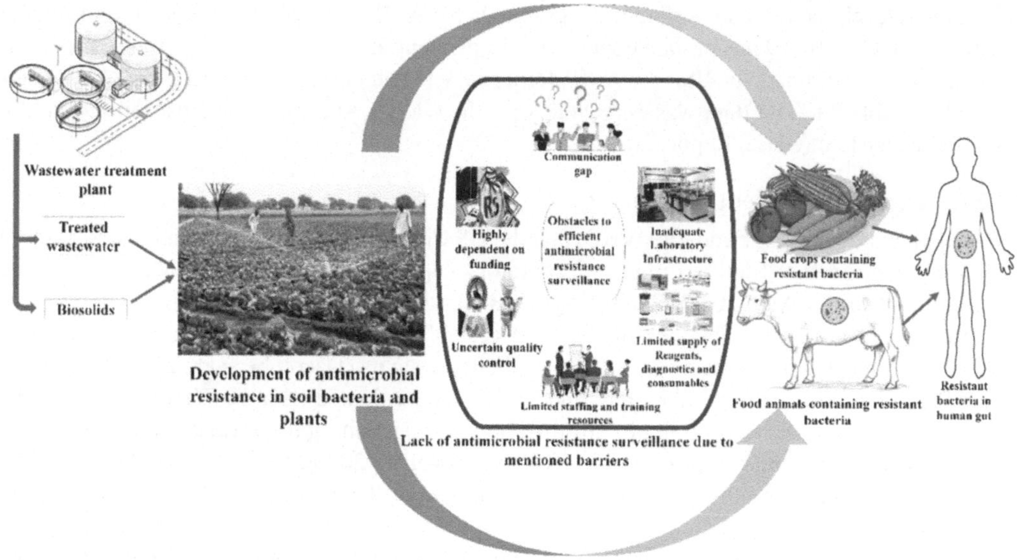

FIGURE 7.3 Antimicrobial resistance transmission and surveillance.

7.8.2.3 Communication Issues

The medical team and laboratory staff's lack of communication is crucial (Tucker & Manyike, 2017). The lack of confidence in laboratory findings is a result of reporting delays and a lack of pertinent information, such as failure to provide minimum inhibitory concentration (MIC), as well as regular diagnostic and reagent shortages (Figure 7.3) (Ombelet et al., 2018). Other factors could be the price of the treatment and the patient's inability to pay for it (Ashley et al., 2019).

7.8.2.4 Limited or Nonexistent Supplies of Reagents, Diagnostics, and Consumables

Diagnostics and reagents that are appropriate for developed countries are needed (Ombelet et al., 2018). High humidity and temperatures can hurt the effectiveness of diagnostics and reagents that need to be kept in a safe and secure cold chain. Electronic devices and other consumables may be harmed by environmental factors (Iskandar et al., 2021). The accuracy of laboratory results may be impacted by item shortages, a lack of or restricted local manufacture, and poor local quality (Barbé et al., 2016a), and the use of inexpensive, subpar reagents and diagnostics (Fitzgibbon & Wallis, 2014). Supply chain problems, such as strict rules for air shipment delivery, may cause protracted delays and call into question the necessity of safe cold chain storage. Products that are susceptible to temperature and humidity may be severely hurt by these supply chain issues (Figure 7.3).

7.8.2.5 Uncertain Quality Control

There is no laboratory guidance for the choice, sampling, or transportation of specimens (Ombelet et al., 2018). The procedure is only partially quality-assured, and quality indicators are not systematically tracked (Frean et al., 2012). It can be difficult to update standard operating procedures and other materials (Fitzgibbon & Wallis, 2014), because of the many languages and cultural barriers that prevent a thorough knowledge of such processes (Barbé et al., 2016b). It is also difficult to implement external quality assurance programmes for all laboratories involved in AMR surveillance (Ashley et al., 2018). In LMICs, automated antimicrobial susceptibility testing does not incorporate the established criteria for bacteria set forth by organizations like the Clinical and Laboratory Standards Institute (CLSI) and the European Committee on Antimicrobial Susceptibility Testing (EUCAST) (CLSI, 2017). Most regulations are only available in English, and are not updated, not followed, or are difficult to understand (Frean et al., 2012) (Figure 7.3). However, in recent years, at the global level, national surveillance programmes for AMR have been implemented in several countries, and COVID-19 has pushed countries to implement these programmes as future preventative measures. A summary of NSP for AMR for a few selected countries around the globe is summarized in Table 7.1. The data was collected from the Global Antimicrobial Resistance and Use Surveillance System (Glass) Report 2021.

7.8.2.6 Highly Dependent on Funding

Agencies like the Fleming Fund, the WHO, and the US Centres for Disease Prevention and Control (CDC) provide external funds to improve laboratory capacity and undertake AMR surveillance programmes (Gandra et al., 2020). Enhancing AMR surveillance in LMICs is the main objective of these funding programmes. The Fleming Fund was established by the UK Department of Health to assist low-income nations in creating AMR surveillance systems (Dacombe et al., 2016) The Global Action Plan on AMR is supported by the fund, which is coordinated with the WHO's Global AMR Surveillance System (O'Neill & Team, 2014b) The Fleming Fund gave various nations a total of £265 million to help increase capacity in LMICs (Gandra et al., 2020). Vietnam, Bangladesh, India, Laos, Nepal, Pakistan, and the Fleming Fund have all received country funding to launch (Gandra et al., 2020).

Global health is seriously threatened by antibiotic resistance, according to surveillance conducted all around the world. To minimize or at least moderate the rate at which resistance is rising, numerous measures are being attempted. Following the identification of the challenges presented by antibiotic resistance, it is now necessary to evaluate the success and effects of these programmes and interventions using both new and current surveillance methods. To effectively tackle the threat of resistance, the timely and focused transmission of surveillance data is crucial.

7.9 APPROACHES TO TACKLE AMR

7.9.1 Global Actions

Health authorities all over the globe, including India, have designed action plans for the management of AMR. For the containment of AMR, it is very important to consider it under the umbrella of the "One Health" concept, which includes mainly four spheres: humans, environment, food, and animals (MoHFW, 2017b). The World Health Assembly (WHA) formulated a Global Action Plan on AMR (GAP-AMR) in May 2014. The Indian Ministry of Health and Family Welfare (MoHFW) initiated the National Action Plan (NAP) for controlling AMR in April 2017 and presented it at the 70th WHA in Geneva in May 2017 (S. G. Kumar et al., 2013). The GAP-AMR proposed five objectives:

- Improve awareness and understanding of AMR.
- Strengthen knowledge through surveillance and research.
- Reduce the incidence of AMR.
- Optimize the use of antimicrobial agents in the health, animal, and food sectors.
- Develop the economic case for sustainable investment that considers all countries' needs and increases investment in new medicines, diagnostic tools, vaccines, and other interventions.

7.9.2 Actions at the International Level (Countrywide)

Along with global actions taken to tackle AMR, strategies have also been implemented by many nations all over the globe to combat antibiotic misuse in health and environment (Table 7.2).

TABLE 7.1
AMR National Surveillance Programmes for Selected Countries

Countries	Australia	Bangladesh	Brazil	Canada	France	India	Italy	Japan	Malaysia	Nepal	Pakistan	South Africa	USA
Population	25.5 million	163.05 million	211.05 million	37.41 million	65.13 million	1.37 billion	59.64 million	128.86 million	31.95 million	28.61 million	216.57 million	58.56 million	329.07 million
World Bank country classification by income	HIC	LMIC	UMIC	HIC	HIC	LMIC	HIC	HIC	UMIC	LMIC	LMIC	UMIC	HIC
GLASS-AMR	Yes	Yes	Yes	Yes	Yes	Yes	Yes	Yes	Yes	Yes	Yes	Yes	Yes
National Coordinating Centre	Established	Established	Established	Established	Established	Established	Established	Establishment in progress	Establishment in progress	Established	Established	Established	Established
National action plan	In place	In place	In place	In place	In place	In place	In place	In place	In place	In place	In place	In place	In place
Number of enrolled National Surveillance Centres	41	9	17	73	2237	71	310	2121	43	41	40	737	48,126
Number of enrolled hospitals	0	0	5	73	991	69	310	2121	42	20	19	535	4035
Inpatient/ outpatient facilities	41 in-outpatient facilities	9 in-outpatient facilities	6 outpatient and 6 in-outpatient facilities	0	1016 outpatient facilities	2 outpatient facilities	0	0	1 outpatient facilities	1 outpatient facilities, 20 in-outpatient facilities	1 outpatient facilities, 20 in-outpatient facilities	200 outpatient facilities, 2 in-outpatient facilities	40,600 outpatient facilities, 3491 in-outpatient facilities
AST standard	EUCAST/ CLSI/Others	CLSI	EUCAST	EUCAST/ CLSI	EUCAST/ CLSI	CLSI	EUCAST	CLSI	CLSI	CLSI	CLSI	EUCAST/CLSI	CLSI
National reference Laboratory	Established	Established	Established	Established	Established	Established	Established	Established	Established	Established	Established	Established	Established
EQA	Provided	Provided	Provided	Provided	Provided	Not provided	Provided	Not Provided	Provided	Provided	Provided	Provided	Provided
Number of Laboratories performing AST	34	8	6	73	Not reported	71	130	2121	0	21	20	79	15,989
EQA provided for bacterial identification	All laboratories	All laboratories	Not provided	All laboratories	All laboratories	Not provided	All laboratories	Not provided	All laboratories	All laboratories	All laboratories	All laboratories	All laboratories

7.9.3 Actions at the Community Level

The following are the different measures aimed at controlling and preventing AMR at the community level that urgently require implementation. In developing countries, irrational antibiotic use is one of the leading causes of AMR, with less than 40% of patients in the public sector and less than 30% of patients in the private sector receiving medical treatment per standard treatment guidelines (Holloway & van Dijk, 2011). Thus, awareness of the rational use of antibiotics is the need of the hour. In developing countries, OTC medications prescribed by minimally qualified and unqualified allied healthcare professionals for common infections are common. The non-prescription use of antimicrobials ranged from as low as 3% in northern European studies to 100% in African studies. Strict regulatory enforcement is urgently needed, particularly for pharmacies dispensing antibiotics (Morgan et al., 2011b; Pokharel & Adhikari, 2020).

7.9.4 Treatment Strategies

High concentrations of ARGs and antibiotics contaminate municipal wastewater, dairy farm soils, and wastewater treatment plants. There are some treatment methods available to reduce the occurrence of AMR in the environment.

7.9.4.1 Anaerobic and Aerobic Treatment Reactors

Aerobic and anaerobic treatment processes are low-energy, eco-friendly, and effective at removing ARGs. In the presence of oxygen, aerobic treatment processes convert organic contaminants to carbon dioxide, water, and biomass. However, anaerobic treatment processes convert organic contaminants to methane, carbon dioxide, and biomass without oxygen (Christgen et al., 2015; Grady Jr et al., 2011). When wastewater was treated in an anaerobic/anoxic/aerobic membrane biological reactor (MBR), the concentration of ARGs in the wastewater diminished in the anaerobic and anoxic effluents, while an increase was observed in the aerobic effluent. As a result, it was concluded that anaerobic and anoxic treatments are far more effective than aerobic treatments in removing ARGs (Du et al., 2015).

7.9.4.2 Constructed Wetlands

Constructed wetlands are small, semi-aquatic ecosystems in which a diverse population of microbial communities thrives and various physicochemical reactions take place. Because of their simplicity, low cost, and effectiveness in removing ARGs, these can be used as ARG treatment plants. It was reported that, in the winter and summer, constructed wetlands removed 14 targeted ARGs at rates of 77.8% and 59.5%, respectively (Fang et al., 2017).

7.9.4.3 Disinfection

Chlorination, ozonation, and UV radiation can disrupt antibiotic-resistant bacteria (ARBs) and ARGs and inactivate harmful microorganisms in wastewater. In a previous study, nine different ARBs and two series of ARGs [ere(A), ere(B), erm(A), erm(B), tet(A), tet(B), tet(M), and tet(O)] were analyzed by RT-PCR after wastewater treatment by chlorination. The quantitative real-time PCR examination and analysis revealed that these genes were removed by 60% and 20%, respectively (Yuan et al., 2015). Results from this study suggest that chlorination was effective in the inactivation of ARBs.

Apart from these treatment methods, there are a few novel methods for tackling AMR in environments mentioned in some studies: coagulation, the use of biochar, and nanomaterials, which need to be studied in detail.

TABLE 7.2
Global Policies and Action Plans to Tackle AMR

Sr. No	Country	Action Plan	Year	Implementing Agency	Objective	Reference
1	India	National Action Plan on Antimicrobial Resistance: India	2016	The Ministry of Health & Family Welfare	Surveillance of antimicrobial resistance in human, animal, food, and environment sectors.	(MoHFW, 2017b)
2	India	National Programme on the Containment of Antimicrobial Resistance	2012	National Centre for Disease Control (NCDC)	Establish a laboratory-based AMRsurveillance system	(MoHFW, 2017b)
3	Canada	Tackling Antimicrobial Resistance and Antimicrobial Use: A Pan-Canadian Framework for Action	2017	Government of Canada	To outline the strategic objectives, outcomes, and opportunities to guide collaborative action on AMR and antimicrobial use (AMU)	(GOC, 2017)
4	Bangladesh	National Action Plan: Antimicrobial Resistance Containment in Bangladesh	2017	Interministerial National Steering Committee (NSC), Ministry of Health and Family Welfare (MoHFW) Bangladesh and	To establish a multi-sectorial approach for planning, coordination, and implementation of antimicrobial resistance containment (ARC) activities	(CDU et al., 2017)
5	Indonesia	National Action Plan on Antimicrobial Resistance: Indonesia	2017	Human Development and Culture, Human Research Development, MoH, Drug and Food Supervisory Agency (BPOM), Ministry of Research and Technology and Higher Education	Reduce the incidence of infection with improved sanitation, hygiene, and infection prevention, and strengthening knowledge and the evidence base through surveillance and research.	(KKRI, 2017)
6	Austria	NAP-AMR: National action plan on antimicrobial resistance	2021	Federal Ministry for Social Affairs, Health, Care and Consumer Protection	To do justice to international developments in combating antimicrobial resistance, which includes human medical and veterinary fields of activity, agriculture, food production, and the environment.	(ECDPC, 2022)

(Continued)

TABLE 7.2 CONTINUED

Sr. No	Country	Action Plan	Year	Implementing Agency	Objective	Reference
7	Belgium	Belgian "one health" national action plan on the fight against antimicrobial resistance	2021	Belgian Federal Public Service	To Establish quantified and measurable targets for reducing antimicrobial use	(FPS Public Health, 2022)
8	Croatia	National Programme for the Control of Antibiotic-Resistant Bacteria	2017	Ministry of Health	To limit the development and spread of antibiotic resistance by raising awareness about the harmfulness of the excessive use of antibiotics among those who prescribe, dispense, and consume antibiotics.	(Holloway & van Dijk, 2011)
9	Denmark	National strategy of Cyprus against microbial resistance to antibiotics	2017	Ministry of Environment and Food of Denmark	Reduction of antibiotic use and the prevention of resistance among humans and animals.	(ECDPC, 2022)
10	Finland	National action plan on antimicrobial resistance	2017	Ministry of Social Affairs and Health	Reduce the incidence of infection through effective sanitation, hygiene, and infection prevention measures	(ECDPC, 2022)
11	France	National strategy for prevention of infections and antibiotic resistance	2022	Ministry of Solidarity and Health	To prevent antibiotic-resistant and multidrug-resistant infections and limit the transmission of antibiotic-resistant bacteria and resistance genes through hygiene measures.	(ECDPC, 2022)
12	Germany	DART 2020 – Fighting antibiotic resistance for the good of both humans and animals	2020	German Federal Government	Strengthening the One Health approach nationally and internationally by recognizing changes in resistance at an early stage, retaining and improving therapy options, breaking chains of infection early and avoiding infections, and supporting research and development	(ECDPC, 2022)

7.10 Role of Artificial Intelligence in Mitigating AMR

Artificial intelligence (AI) has broad applications across various fields and has been implemented in AMR management (Chowdhury et al., 2020; Rabaan et al., 2022; Voermans et al., 2019). AI has accelerated the development of new antibiotics as it is useful for identifying potential antimicrobial molecules, testing antimicrobial susceptibility, and optimizing drug combinations. Due to AI's automation and efficiency, it can play an influential role in mitigating AMR (Nava Lara et al., 2019; Weinstein et al., 2017). AMR is typically diagnosed by conventional antibiotic susceptibility testing (AST) and whole-genome sequencing (WGS). Although AST is a classical method, it is less efficient and does not explain the AMR mechanism. In contrast, WGS is a rapid and accurate method requiring extensive and high-dimensional datasets (Blair et al., 2015; Horne et al., 2013). When AI is combined with AST and WGS, rapid results are obtained (Inglis et al., 2020).

7.10.1 AI Algorithms in Combating AMR

Decision trees (DT), random forests (RF), artificial neural networks (ANN), Naive Bayes (NB), and support vector machines (SVM) are some of the AI algorithms used for AMR. The NB algorithm was used to monitor AMR, extract resistant patterns, and determine the critical factor of resistance to estimate the probabilities of ineffective treatment due to AMR (Choisy et al., 2019; Gerontini et al., 2011; Rezaei-Hachesu et al., 2018). The decision tree model is commonly used to allocate healthcare resources more appropriately by estimating the burden of AMR (Naylor et al., 2017, 2018). Previously, the decision tree model was used to estimate healthcare utilization and cost for AMR and to guide antibiotic use, resulting in lower antibiotic doses and shorter treatment periods (Reynolds et al., 2014; Schuetz et al., 2012). Random forest is an ensemble algorithm for predicting antibiotic combinations. Previously, chemogenomics data and molecular fingerprints were used to predict an effective antibiotic combination using a random forest algorithm (Chandrasekaran et al., 2016; Mason et al., 2017) The SVM is a binary classification model used to predict AMR phenotypes, such as "resistant" or "susceptible". SVM is more than 90% accurate in testing resistance to tetracycline, ampicillin, sulfisoxazole, trimethoprim, and enrofloxacin (Table 7.3) (Her & Wu, 2018; Liu et al., 2020).

7.11 THE INDIAN SCENARIO

Antimicrobial resistance is a prominent global health threat that has a devastating effect on public health in terms of mortality and economic loss. AMR is one of the Sustainable Development Goals (SDGs) and has been listed among the top ten global public health threats facing humanity by the WHO (WHO, 2021). Despite the efforts that have been initiated by Indian health authorities to combat it, AMR continues to challenge the health system.

7.11.1 Status of AMR in India

AMR is one of the primary causes of the rising burden of bacterial diseases in India. In the past decade, the rate of AMR has increased disproportionately in humans, animals, and the environment. AMR in both Gram-positive and Gram-negative bacteria results in the emergence of drug-resistant strains. The emergence of multi-drug-resistant strains of pathogens has evolved due to antibiotic resistance. Examples include multi-drug-resistant tuberculosis caused by Mycobacterium's extensive drug-resistant strain. *Pseudomonas aeruginosa*, *Enterococcus* sp., *Staphylococcus aureus*, *Escherichia coli*, *Neisseria gonorrhoeae*, and *Klebsiella pneumoniae* are a few prominent bacteria reported in India to develop antibiotic resistance (Aggarwal et al., 2018; WHO, 2022). There are different mechanisms by which bacteria can develop antibiotic resistance. Some novel mechanisms have been reported in India. Examples include vancomycin-resistant *Enterococci*,

Metallo-β-lactamases in Gram-negative bacteria, carbapenems and colistin-resistant bacteria, and AmpC-mediated drug resistance in *Enterobacteriaceae* (Santajit & Indrawattana, 2016). The emergence of "superbugs" poses a significant challenge in front of the medical system because only high-end antibiotics can treat these multi-drug-resistant pathogens. "ESKAPE" (*Enterococcus faecium, Staphylococcus aureus, Klebsiella pneumoniae, Acinetobacter baumannii, Pseudomonas aeruginosa*, and *Enterobacter* species) group are highly virulent bacteria that have acquired the status of superbugs. A research team from the Indian Veterinary Research Institute (IVRI) detected superbugs, including ESKAPE, from Bareilly water as another cause for alarm (Agarwal, 2019). Effluents from pharmaceutical industries and hospitals are the primary sources of antibiotic-resistant genes. High-throughput metagenomic sequencing of microbes in the water of the Ganges revealed that the river harbours a broad spectrum of ARGs. Examples include β-lactam, multidrug/efflux, and elfamycin (Gandra et al., 2017; Reddy & Dubey, 2019). Indian institutions have published 2152 studies on AMR, out of which 1040 (48.3%) were on humans, 90 (4.2%) on the environment, while only 70 (3.3%) on animals, and 11 (0.5%) on One Health (Azam et al., 2016; Gandra et al., 2017) Extended-spectrum β-lactamase (ESBL) bacteria and ARGs genes like *bla NDM-1* and *bla* OXA48 were detected from the north Indian rivers (Ahammad et al., 2014). Bacteria resistant to third-generation cephalosporins were isolated from the Cauvery River in Karnataka (Skariyachan et al., 2015). In central India, the drinking water has been reported to have a 17% rate of *E. coli* resistance to third-generation cephalosporins (S. Kumar et al., 2013), similarly 7% in Kashmir (Rather et al., 2013), 50% in Sikkim (Poonia et al., 2014), and 100% in Hyderabad (Lübbert et al., 2017).

7.11.2 AMR Concerns in India

India is considered one of the epicentres of AMR. The advent of recent multi-drug-resistant (MDR) species poses significant diagnostic and therapeutic challenges to the medical system. The other factors contributing to the increase of AMR in India are poverty, population, malnutrition, and lack of knowledge regarding infectious diseases and the use of antibiotics (Chaudhry & Tomar, 2017; Laxminarayan et al., 2016; Swaminathan et al., 2017). In many cases, limited access to medical facilities precludes people from seeking medical advice, which leads to the self-prescription of medication without any knowledge regarding the uses of drugs. On the other hand, many people receive broad-spectrum antibiotics due to a lack of proper diagnosis of infectious agents causing illness (Morgan et al., 2011a).

India is home to one of the world's largest pharmaceutical production industries (Laxminarayan & Chaudhury, 2016). These pharmaceutical firms produce a significant amount of waste; furthermore, due to the lack of strict supervisory guidelines, this waste enters water bodies and continues to be a source of AMR in the environment (Lundborg & Tamhankar, 2017). Since India is a farming-based economy, another vital source of AMR in the environment is the uncontrolled use of antimicrobial agents such as pesticides and insecticides. However, there is no confirmed evidence for this claim (Forsberg et al., 2014).

7.11.3 Environmental AMR Drivers in India

Environmental routes for AMR's emergence are substantially higher in developing countries, including India. The factors contributing to the development of environmental AMR in India are discussed here.

7.11.3.1 Municipal Wastewater

It was reported that municipal wastewater is a dumping ground for 30–90% of excreted antibiotics via human faeces. Of the total sewage water, only 20–30% is treated in the treatment plant before being released into bodies of water, which is certainly not enough to eliminate the drug-resistant species. The rate of vancomycin-resistant *Enterococci*, a normal commensal human gut bacterium, was reported to be between 22–100% in the Gomti River (Lata et al., 2016).

7.11.3.2 Hospital Waste

Hospitals and all other healthcare settings are significant sources of antimicrobial waste, either directly through patient secretion or indirectly through unused medicine. Fluoroquinolone, sulphonamide, and tinidazole residues were found in the hospital effluents in India. The effluent water of hospitals is the richest source of AMR genes and resistant bacteria, as many antibiotics are used to treat patients in hospitals. The high concentration of antimicrobials released by hospitals causes bacterial strains to undergo genotoxic alterations and modifications (Diwan et al., 2010; Mutiyar & Mittal, 2014). Hospital effluent should undergo proper treatment before disposal, but less than 45% of healthcare units in India have efficient wastewater treatment facilities.

7.11.3.3 Pharmaceutical Wastewater

India is among the world's largest manufacturers of pharmaceuticals and, as a result, it generates a sizeable amount of pharmaceutical waste. Ciprofloxacin concentrations of 28 mg/L and 31 mg/L for two consecutive days were found in the effluent of one Indian pharmaceutical plant. In India, several kilogrammes of antibiotics are dumped into the wastewater daily (Larsson et al., 2007; Rehman et al., 2015). It was reported that β-lactam drugs degrade relatively faster than antimicrobials like sulphonamides and fluoroquinolones, which leave behind stable residues in the environment. These drug classes are the main contributors to AMR pollutants in nearby water bodies and are widely found in the wastewater of Indian pharmaceutical industries (Fick et al., 2009). Even though wastewater is also produced by the pharmaceutical industry in developed nations, the overall risk of AMR in the environment is low because wastewater is properly treated before being released into the environment. On the contrary, the risk of AMR is high in India due to a lack of efficient wastewater treatment facilities (Chereau et al., 2017). Additionally, due to the high cost, monitoring antimicrobials in industrial effluent is given less attention in India (Mutiyar & Mittal, 2014).

7.11.3.4 Aquaculture

Aquaculture is one of the booming sectors of the food industry in India. Because of the unchecked use of antimicrobials to meet the rising demand for seafood, aquaculture is turning into a new environmental source of AMR. Moreover, the possibility of AMR spreading through international waters is greater in the aquatic environment than in soil and wastewater, which are geographically more confined. However, there is no clear evidence regarding the number of antimicrobial residues reported from aquaculture sources in India. (Henriksson et al., 2015; Swapna K.M et al., 2012).

7.11.4 Strategies for Tackling AMR in India

The South East Asian Regional Office (SEARO) held a meeting in Jaipur, India, in 2011, at which health ministers from respective countries adopted the "Jaipur Declaration" to combat AMR (WHO, 2011). In 2014, the WHO designated AMR as a flagship priority area for SEARO, and the Indian Medical Association announced an AMR awareness programme to train physicians and build awareness among the public (Jha, 2014; WHO-ROSEA, 2017). The Indian government prepared a National Action Plan to address AMR in the environment; moreover, the Central Drugs Standard Control Organization implemented Schedule H1 to limit the illegal sale of drugs over the counter (Laxminarayan & Chaudhury, 2016; MoHFW, 2017b). In 2013 and 2014, the Indian Council of Medical Research and the National Centre for Disease Control launched an AMR surveillance network, respectively, as well as a systematic assessment of India's current infection prevention and control (IPC) practices (IPC) in collaboration with the Centers for Disease Control and Prevention (CDC) in the United States. The National Health Systems Resource Centre of the National Health Mission and the National Accreditation Board regulate IPC practices and aim to develop new

TABLE 7.3
AI Models Used in the MIC Prediction of Antibiotic Resistance

Sr. No	Pathogen	AI Method	Antibiotics	MIC prediction Accuracy%	Reference
1	*Salmonella spp.*	Machine Learning	Chloramphenicol	99	(Nguyen et al., 2019)
			ceftiofur	99	
			gentamicin	91	
			tetracycline	90	
2	*Neisseria gonorrhoeae*	Machine Learning	Cefixime	91	(Eyre et al., 2017)
			Penicillin	93	
			Azithromycin	94	
			Ciprofloxacin	96	
			Tetracycline	93	
3	*Klebsiella pneumoniae*	Machine Learning	Amikacin	97	(Nguyen et al., 2018)
			Ampicillin	100	
			Ampicillin	99	
			Aztreonam	89	
			Cefazolin	96	
			Cefepime	61	
			Cefoxitin	90	
			Ceftazidime	92	
			Ceftriaxone	89	
			Cefuroxime sodium	99	

guidelines for preventing hospital-acquired infections (Laxminarayan & Chaudhury, 2016). The National Health Policy guidelines for 2017 limited the use of antibiotics as OTC medications and prohibited the use of antibiotics as livestock growth promoters. It also performs pharmacovigilance work, such as prescription audits in hospitals and community settings (MoHFW, 2017a).

7.12 CONCLUSION AND FUTURE PROSPECTS

The studies and surveys indicate that AMR can be the next big pandemic in the coming years. The ability of microbes to cause several human and animal diseases and the resistance of these microbes against antibiotics that are used to treat the disease are keeping scientists and doctors from developing strategies to treat the patients. The rapid transmission and evolution of AMR is a serious concern for mankind. Therefore, comprehensive data on population-based surveillance of AMR is very important. However, inadequate laboratory capability, poor governance, lack of health and associated information systems, and limited resources for AMR surveillance in low- and middle-income countries impact the global fight against AMR. Also, political situations and social beliefs increase communicable diseases and associated economic burdens. Although some countries are attempting to support data surveillance in association with GLASS, the availability of fragmented data and poor health infrastructure limits them from developing policies and allocating funds toward surveillance systems. However, it needs to be understood that low-quality data on AMR will result in skyrocketing rates of antimicrobial resistance that will later turn into a pandemic. The COVID-19 pandemic has helped many countries, including India, strengthen their surveillance systems. Thus, taking inspiration from these, similar strategies can be developed in tracking and mapping the spread of AMR, helping early detection of outbreaks, and also devising national health policies to tackle AMR or any future pandemic. However, a one-size-fits-all policy cannot be applied at the global level. Thus, effective yet economical surveillance and treatment action policies, sensing the peculiarities of the respective country and aligning them to regional, national, and global efforts, are required. The developed and high-income developing countries have a key role in sharing their research and infrastructure to train and develop surveillance and treatment infrastructure in low- and middle-income countries to prevent any future pandemic.

CONFLICT OF INTEREST

All the authors approve of the submission and do not have any conflicts of interest to declare.

ACKNOWLEDGMENT

BK wishes to acknowledge the DST-SERB National Post-Doctoral Fellowship (PDF/2022/001781) for financial support.

REFERENCES

Agarwal, P. (2019, October 22). *Bareilly Water Samples Spark 'Superbug' Scare.* https://timesofindia.indiatimes.com/city/bareilly/more-than-half-of-community-water-sample-in-bareilly-detected-with-superbug-ivri-research/articleshow/71694601.cms

Aggarwal, R., Rastogi, N., Mathur, P., Soni, K., Kumar, S., Gupta, A., & Sagar, S. (2018). Colistin-resistant Klebsiella pneumoniae in surgical polytrauma intensive care unit of level-1 trauma center: First case series from trauma patients in India. *Indian Journal of Critical Care Medicine*, *22*(2), 103–106. https://doi.org/10.4103/ijccm.IJCCM_430_17

Ahammad, Z. S., Sreekrishnan, T. R., Hands, C. L., Knapp, C. W., & Graham, D. W. (2014). Increased waterborne bla NDM-1 resistance gene abundances associated with seasonal human pilgrimages to the Upper Ganges River. *Environmental Science and Technology*, *48*(5), 3014–3020. https://doi.org/10.1021/es405348h

Ahmed, A. M., & Gareib, M. M. (2016). Detection of some antibiotics residues in chicken meat and chicken luncheon. *Environmental Health*, *2*(2), 315–323.

Ashley, E. A., Recht, J., Chua, A., Dance, D., Dhorda, M., Thomas, N. v., Ranganathan, N., Turner, P., Guerin, P. J., White, N. J., & Day, N. P. (2018). An inventory of supranational antimicrobial resistance surveillance networks involving low- and middle-income countries since 2000. *Journal of Antimicrobial Chemotherapy*, *73*(7), 1737–1749. https://doi.org/10.1093/jac/dky026

Ashley, E. A., Shetty, N., Patel, J., van Doorn, R., Limmathurotsakul, D., Feasey, N. A., Okeke, I. N., & Peacock, S. J. (2019). Harnessing alternative sources of antimicrobial resistance data to support surveillance in low-resource settings. *Journal of Antimicrobial Chemotherapy*, *74*(3), 541–546. https://doi.org/10.1093/jac/dky487

Azam, M., Jan, A. T., & Haq, Q. M. R. (2016). blaCTX-M-152, a novel variant of CTX-M-group-25, identified in a study performed on the prevalence of multidrug resistance among natural inhabitants of river Yamuna, India. *Frontiers in Microbiology*, *7*(FEB). https://doi.org/10.3389/fmicb.2016.00176

Barbé, B., Verdonck, K., El-Safi, S., Khanal, B., Teav, S., Lilo Kalo, J. R., Ravinetto, R., Chappuis, F., Boelaert, M., & Jacobs, J. (2016a). Rapid diagnostic tests for neglected infectious diseases: Case study highlights need for customer awareness and postmarket surveillance. *PLoS Neglected Tropical Diseases*, *10*(11). https://doi.org/10.1371/journal.pntd.0004655

Bate, R., Tren, R., Mooney, L., Hess, K., Mitra, B., Debroy, B., & Attaran, A. (2009). Pilot study of essential drug quality in two major cities in India. *PLoS ONE*, *4*(6). https://doi.org/10.1371/journal.pone.0006003

Bearson, B. L., & Brunelle, B. W. (2015). Fluoroquinolone induction of phage-mediated gene transfer in multidrug-resistant Salmonella. *International Journal of Antimicrobial Agents*, *46*(2), 201–204. https://doi.org/10.1016/j.ijantimicag.2015.04.008

Bebell, L. M., & Muiru, A. N. (2014). Antibiotic use and emerging resistance: How can resource-limited countries turn the tide? In *Global Heart* (Vol. 9, Issue 3, pp. 347–358). Elsevier B.V. https://doi.org/10.1016/j.gheart.2014.08.009

Bengtsson-Palme, J., Jonsson, V., & Heß, S. (2021). What is the role of the environment in the emergence of novel antibiotic resistance genes? A modeling approach. *Environmental Science and Technology*, *55*(23), 15734–15743. https://doi.org/10.1021/acs.est.1c02977

Bengtsson-Palme, J., & Larsson, D. G. J. (2016). Concentrations of antibiotics predicted to select for resistant bacteria: Proposed limits for environmental regulation. *Environment International*, *86*, 140–149. https://doi.org/10.1016/j.envint.2015.10.015

Beyene, T. (2015). Veterinary drug residues in food-animal products: Its risk factors and potential effects on public health. *Journal of Veterinary Science & Technology*, *07*(01). https://doi.org/10.4172/2157-7579.1000285

Blair, J. M. A., Webber, M. A., Baylay, A. J., Ogbolu, D. O., & Piddock, L. J. v. (2015). Molecular mechanisms of antibiotic resistance. *Nature Reviews Microbiology*, *13*(1), 42–51. https://doi.org/10.1038/nrmicro3380

Bloom, B., Atun, R., Cohen, T., Dye, C., Fraser, H., Gomez, G. B., Knight, G., Murray, M., Nardell, E., Rubin, E., Salomon, J., Vassall, A., Volchenkov, G., White, R., Wilson, D., & Yadav, P. (2017). Tuberculosis. In K.K. Holmes, S. Bertozi, R.B. Bloom, & P. Jha (Eds.), *Major Infectious Diseases* (Vol. 6, pp. 233–314), The International Bank for Reconstruction and Development/The World Bank. Washington (DC). https://doi.org/10.1596/978-1-4648-0524-0_ch11.

Boerma, T., & Mathers, C. D. (2015). The World Health Organization and global health estimates: Improving collaboration and capacity. In *BMC Medicine* (Vol. 13, Issue 1). BioMed Central Ltd. https://doi.org/10.1186/s12916-015-0286-7

Bokhary, H., Pangesti, K. N. A., Rashid, H., Abd El Ghany, M., & Hill-Cawthorne, G. A. (2021). Travel-related antimicrobial resistance: A systematic review. In *Tropical Medicine and Infectious Disease* (Vol. 6, Issue 1). MDPI AG. https://doi.org/10.3390/tropicalmed6010011

Brigulla, M., & Wackernagel, W. (2010). Molecular aspects of gene transfer and foreign DNA acquisition in prokaryotes with regard to safety issues. In *Applied Microbiology and Biotechnology* (Vol. 86, Issue 4, pp. 1027–1041). https://doi.org/10.1007/s00253-010-2489-3

Bruno, A., & Carolissen Mackay, V. (2012). Antimicrobial resistance and the activities of the Codex Alimentarius Commission. *Revue Scientifique et Technique (International Office of Epizootics)*, *1*, 317–323.

Canteón, R. (2009). Antibiotic resistance genes from the environment: A perspective through newly identified antibiotic resistance mechanisms in the clinical setting. *Clinical Microbiology and Infection*, *15*, 20–25. https://doi.org/10.1111/j.1469-0691.2008.02679.x

CDC. (2022, October 5). *About Antimicrobial Resistance*. Centers for Disease Control and Prevention, National Center for Emerging and Zoonotic Infectious Diseases (NCEZID), Division of Healthcare Quality Promotion (DHQP). https://www.cdc.gov/drugresistance/about.html

CDU, CDC, & DGHS. (2017, April 23). *Bangladesh: Antimicrobial Resistance Containment in Bangladesh 2017-2022*. https://cdn.who.int/media/docs/default-source/antimicrobial-resistance/amr-spc-npm/nap-library/antimicrobial-resistance-containment-in-bangladesh-2017-2022.pdf?sfvrsn=bfa46b_3&download=true

Chandrasekaran, S., Cokol-Cakmak, M., Sahin, N., Yilancioglu, K., Kazan, H., Collins, J. J., & Cokol, M. (2016). Chemogenomics and orthology-based design of antibiotic combination therapies. *Molecular Systems Biology*, *12*(5), 872. https://doi.org/https://doi.org/10.15252/msb.20156777

Chaudhry, D., & Tomar, P. (2017). Antimicrobial resistance: the next BIG pandemic. *International Journal Community Medicine Public Health*, *4*(8), 2632–2636.

Chen, J., Quiles-Puchalt, N., Chiang, N., Bacigalupe, R., Fillol-Salom, A., Su, M., Chee, J., Fitzgerald, J. R., & Penadés, J. R. (2018). Genome hypermobility by lateral transduction. In *Science* (Vol. 362). https://www.science.org

Chereau, F., Opatowski, L., Tourdjman, M., & Vong, S. (2017). Risk assessment for antibiotic resistance in South East Asia. *BMJ (Online)*, *358*, 2–8. https://doi.org/10.1136/bmj.j3393

Choisy, M., van Cuong, N., Bao, T. D., Kiet, B. T., Hien, B. V., Thu, H. V., Chansiripornchai, N., Setyawan, E., Thwaites, G., Rushton, J., & Carrique-Mas, J. (2019). Assessing antimicrobial misuse in small-scale chicken farms in Vietnam from an observational study. *BMC Veterinary Research*, *15*(1). https://doi.org/10.1186/s12917-019-1947-0

Chowdhury, A. S., Lofgren, E. T., Moehring, R. W., & Broschat, S. L. (2020). Identifying predictors of antimicrobial exposure in hospitalized patients using a machine learning approach. *Journal of Applied Microbiology*, *128*(3), 688–696. https://doi.org/https://doi.org/10.1111/jam.14499

Christgen, B., Yang, Y., Ahammad, S. Z., Li, B., Rodriquez, D. C., Zhang, T., & Graham, D. W. (2015). Metagenomics shows that low-energy anaerobic-aerobic treatment reactors reduce antibiotic resistance gene levels from domestic wastewater. *Environmental Science and Technology*, *49*(4), 2577–2584. https://doi.org/10.1021/es505521w

CLSI. (2017). *Performance Standards for Antimicrobial Susceptibility Testing* (Vol. 27). https://file.qums.ac.ir/repository/mmrc/clsi%202017.pdf

Cohen, M. (1997). Epidemiological factors influencing the emergence of antimicrobial resistance. *Ciba Found Symp, 207,* 223–231.

Dacombe, R., Pulford, J., Wallis, S., Bhardwaj, M., & Bates, I. (2016). *Fleming Fund: Supporting Surveillance Capacity for Antimicrobial Resistance Regional Networks and Educational Resources,* Capacity Research Unit, Liverpool School of Tropical Medicine, UK.

Dahal, R., Upadhyay, A., & Ewald, B. (2017). One health in South Asia and its challenges in implementation from stakeholder perspective. *Veterinary Record, 181*(23), 626. https://doi.org/https://doi.org/10.1136/vr.104189

Dar, O. A., Hasan, R., Schlundt, J., Harbarth, S., Caleo, G., Dar, F. K., Littmann, J., Rweyemamu, M., Buckley, E. J., Shahid, M., Kock, R., Li, H. L., Giha, H., Khan, M., So, A. D., Bindayna, K. M., Kessel, A., Pedersen, H. B., Permanand, G., … Heymann, D. L. (2016). Exploring the evidence base for national and regional policy interventions to combat resistance. *The Lancet, 387*(10015), 285–295. https://doi.org/https://doi.org/10.1016/S0140-6736(15)00520-6

D'Atri, F., Arthur, J., Blix, H. S., Hicks, L. A., Plachouras, D., Monnet, D. L., Strauß, R., Ingenbleek, A., Christova, I., Andrašević, A. T., Žemličkova, H., Moloney, C., Märtin, P., Jalava, J., Cavalié, P., Eckmanns, T., Kontopidou, F., Matuz, M., Gudnason, T., … Reilly, J. (2019). Targets for the reduction of antibiotic use in humans in the Transatlantic Taskforce on Antimicrobial Resistance (TATFAR) partner countries. In *Eurosurveillance* (Vol. 24, Issue 28). European Centre for Disease Prevention and Control (ECDC). https://doi.org/10.2807/1560-7917.ES.2019.24.28.1800339

Davies, J., & Davies, D. (2010). Origins and evolution of antibiotic resistance. In *Microbiology and Molecular Biology Reviews* (Vol. 74, Issue 3, pp. 417–433). https://doi.org/10.1128/mmbr.00016-10

D'Costa, V. M., Griffiths, E., & Wright, G. D. (2007). Expanding the soil antibiotic resistome: exploring environmental diversity. *Current Opinion in Microbiology, 10*(5), 481–489. https://doi.org/https://doi.org/10.1016/j.mib.2007.08.009

de Kraker, M. E. A., Stewardson, A. J., & Harbarth, S. (2016). Will 10 million people die a year due to antimicrobial Resistance by 2050? *PLoS Medicine, 13*(11). https://doi.org/10.1371/journal.pmed.1002184

Di, K. N., Tay, S. T., Ponnampalavanar, S. S. L. S., Pham, D. T., & Wong, L. P. (2022). Socio-demographic factors associated with antibiotics and antibiotics Resistance knowledge and practices in vietnam: A cross-sectional survey. *Antibiotics, 11*(4). https://doi.org/10.3390/antibiotics11040471

Diwan, V., Tamhankar, A. J., Khandal, R. K., Sen, S., Aggarwal, M., Marothi, Y., Iyer, R. v, Sundblad-Tonderski, K., & Stålsby-Lundborg, C. (2010). Antibiotics and antibiotic-resistant bacteria in waters associated with a hospital in Ujjain, India. In *BMC Public Health* (Vol. 10). http://www.biomedcentral.com/1471-2458/10/414

DOH-UK, VMD-UK, & DEFRA-UK. (2013). *UK Five Year Antimicrobial Resistance Strategy 2013 to 2018.* www.nationalarchives.gov.uk/doc/open-government-licence/

Du, J., Geng, J., Ren, H., Ding, L., Xu, K., & Zhang, Y. (2015). Variation of antibiotic resistance genes in municipal wastewater treatment plant with A(2)O-MBR system. *Environmental science and pollution research international,* 22(5), 3715–3726. https://doi.org/10.1007/s11356-014-3552-x

ECDC. (2018). *33000 People Die Every Year Due to Infections with Antibiotic-resistant Bacteria.* European Centre for Disease Prevention and Control. https://www.ecdc.europa.eu/en/news-events/33000-people-die-every-year-due-infections-antibiotic-resistant-bacteria

ECDPC. (2022). *Strategies and Action Plans on Antimicrobial Resistance.* https://www.ecdc.europa.eu/en/publications-data/directory-guidance-prevention-and-control/antimicrobial-resistance-strategies

Eyre, D. W., Silva, D. de, Cole, K., Peters, J., Cole, M. J., Grad, Y. H., Demczuk, W., Martin, I., Mulvey, M. R., Crook, D. W., Walker, A. S., Peto, T. E. A., & Paul, J. (2017). WGS to predict antibiotic MICs for Neisseria gonorrhoeae. *Journal of Antimicrobial Chemotherapy, 72*(7), 1937–1947. https://doi.org/10.1093/jac/dkx067

Fang, H., Zhang, Q., Nie, X., Chen, B., Xiao, Y., Zhou, Q., Liao, W., & Liang, X. (2017). Occurrence and elimination of antibiotic resistance genes in a long-term operation integrated surface flow constructed wetland. *Chemosphere, 173,* 99–106. https://doi.org/https://doi.org/10.1016/j.chemosphere.2017.01.027

FAO, OIE, & WHO. (2018). *Monitoring Global Progress on Addressing Antimicrobial Resistance Analysis Report of the Second Round of Results of AMR Country Self-assessment Survey 2018.* https://www.who.int/publications/i/item/monitoring-global-progress-on-addressing-antimicrobial-resistance

Fick, J., Söderström, H., Lindberg, R. H., Phan, C., Tysklind, M., & Larsson, D. G. J. (2009). Contamination of surface, ground, and drinking water from pharmaceutical production. *Environmental Toxicology and Chemistry, 28*(12), 2522–2527. https://doi.org/https://doi.org/10.1897/09-073.1

Fitzgibbon, J. E., & Wallis, C. L. (2014). Laboratory challenges conducting international clinical research in resource-limited settings. *Journal of Acquired Immune Deficiency Syndromes, 65*(Suppl. 1). https://doi.org/10.1097/QAI.0000000000000038

Fletcher, S. (2015). Understanding the contribution of environmental factors in the spread of antimicrobial resistance. In *Environmental Health and Preventive Medicine* (Vol. 20, Issue 4, pp. 243–252). Springer Tokyo. https://doi.org/10.1007/s12199-015-0468-0

Forsberg, K. J., Patel, S., Gibson, M. K., Lauber, C. L., Knight, R., Fierer, N., & Dantas, G. (2014). Bacterial phylogeny structures soil resistomes across habitats. *Nature, 509*(7502), 612–616. https://doi.org/10.1038/nature13377

FPS Public Health. (2022). *Belgian "One Health" National Action Plan on the Fight against Antimicrobial Resistance 2020-2024.* https://www.health.belgium.be/en/belgian-one-health-national-action-plan-fight-against-antimicrobial-resistance-2020-2024

Frean, J., Perovic, O., Fensham, V., Mccarthy, K., von Gottberg, A., de Gouveia, L., Poonsamy, B., Dini, L., Rossouw, J., Keddy, K., Wondimagegnehu, A., Yahaya, A., Pierson, A., Dolmazon, V., Cognat, S., & Ndihokubwayo, J. B. (2012). Evaluación externa de la calidad de los laboratorios nacionales de la sanidad pública en áfrica, 2002-2009. *Bulletin of the World Health Organization, 90*(3). https://doi.org/10.2471/BLT.11.091876

Gandra, S., Alvarez-Uria, G., Turner, P., Joshi, J., Limmathurotsakul, D., & Rogier Van Doorn, H. (2020). *Antimicrobial Resistance Surveillance in Low-and Middle-Income Countries: Progress and Challenges in Eight South Asian and Southeast Asian Countries.* https://doi.org/10.1128/CMR

Gandra, S., Joshi, J., Trett, A., Sankhil Lamkang, A., & Laxminarayan, R. (2017). *Scoping Report on Antimicrobial Resistance in India,* Washington, DC: Center for Disease Dynamics, Economics & Policy.

Gandra, S., Miller-Petrie, Pant, S., Gandra, S., Levinson, J., Barter, D., White, A., & Laxminarayan, R. (2015). The State of the World. *CDDEP: Washington, D.C.* , *1*(1), 1–85. https://www.researchgate.net/publication/308721773

Gerontini, M., Vazirgiannis, M., Vatopoulos, A. C., & Polemis, M. (2011). Predictions in antibiotics resistance and nosocomial infections monitoring. *Proceedings - IEEE Symposium on Computer-Based Medical Systems.* https://doi.org/10.1109/CBMS.2011.5999112

GOC. (2017). *Tackling Antimicrobial Resistance and Antimicrobial Use_ A Pan-Canadian Framework for Action - Canada.ca.* https://www.canada.ca/en/health-canada/services/publications/drugs-health-products/tackling-antimicrobial-resistance-use-pan-canadian-framework-action.html

Grady Jr, C. P. L., Daigger, G. T., Love, N. G., & Filipe, C. D. M. (2011). *Biological Wastewater Treatment.* CRC press.

Haaber, J., Leisner, J. J., Cohn, M. T., Catalan-Moreno, A., Nielsen, J. B., Westh, H., Penadés, J. R., & Ingmer, H. (2016). Bacterial viruses enable their host to acquire antibiotic resistance genes from neighbouring cells. *Nature Communications, 7.* https://doi.org/10.1038/ncomms13333

Hanna, N., Sun, P., Sun, Q., Li, X., Yang, X., Ji, X., Zou, H., Ottoson, J., Nilsson, L. E., Berglund, B., Dyar, O. J., Tamhankar, A. J., & Stålsby Lundborg, C. (2018). Presence of antibiotic residues in various environmental compartments of Shandong province in eastern China: Its potential for resistance development and ecological and human risk. *Environment International, 114,* 131–142. https://doi.org/10.1016/j.envint.2018.02.003

He, T., Wang, R., Liu, D., Walsh, T. R., Zhang, R., Lv, Y., Ke, Y., Ji, Q., Wei, R., Liu, Z., Shen, Y., Wang, G., Sun, L., Lei, L., Lv, Z., Li, Y., Pang, M., Wang, L., Sun, Q., … Wang, Y. (2019). Emergence of plasmid-mediated high-level tigecycline resistance genes in animals and humans. *Nature Microbiology, 4*(9), 1450–1456. https://doi.org/10.1038/s41564-019-0445-2

Henriksson, P. J. G., Troell, M., & Rico, A. (2015). Antimicrobial use in aquaculture: Some complementing facts. In *Proceedings of the National Academy of Sciences of the United States of America* (Vol. 112, Issue 26, p. E3317). National Academy of Sciences. https://doi.org/10.1073/pnas.1508952112

Her, H. L., & Wu, Y. W. (2018). A pan-genome-based machine learning approach for predicting antimicrobial resistance activities of the Escherichia coli strains. *Bioinformatics, 34*(13), i89–i95. https://doi.org/10.1093/bioinformatics/bty276

Hobman, J. L., & Crossman, L. C. (2015). Bacterial antimicrobial metal ion resistance. In *Journal of Medical Microbiology* (Vol. 64, pp. 471–497). Microbiology Society. https://doi.org/10.1099/jmm.0.023036-0

Holloway, K., & van Dijk, L. (2011). *The World Medicines Situation 2011, Rational Use of Medicines* (3rd ed., K. Holloway & L. van Dijk, Eds., Vol. 3). World Health Organization.

Horne, D. J., Pinto, L. M., Arentz, M., Grace Lin, S. Y., Desmond, E., Flores, L. L., Steingart, K. R., & Minion, J. (2013). Diagnostic accuracy and reproducibility of WHO-endorsed phenotypic drug susceptibility testing methods for first-line and second-line antituberculosis drugs. In *Journal of Clinical Microbiology* (Vol. 51, Issue 2, pp. 393–401). https://doi.org/10.1128/JCM.02724-12

Hubbard, A. T. M., Jafari, N. V., Feasey, N., Rohn, J. L., & Roberts, A. P. (2019, August). Effect of environment on the evolutionary trajectories and growth characteristics of antibiotic-resistant escherichia coli mutants. *Frontiers in Microbiology, 10.* https://doi.org/10.3389/fmicb.2019.02001

Ibrahim, N. K. (2020). Epidemiologic surveillance for controlling Covid-19 pandemic: Types, challenges and implications. In *Journal of Infection and Public Health* (Vol. 13, Issue 11, pp. 1630–1638). Elsevier Ltd. https://doi.org/10.1016/j.jiph.2020.07.019

Inglis, T. J. J., Paton, T. F., Kopczyk, M. K., Mulroney, K. T., & Carson, C. F. (2020). Same-day antimicrobial susceptibility test using acoustic-enhanced flow cytometry visualized with supervised machine learning. *Journal of Medical Microbiology, 69*(5), 657–669. https://doi.org/10.1099/jmm.0.001092

Iskandar, K., Molinier, L., Hallit, S., Sartelli, M., Hardcastle, T. C., Haque, M., Lugova, H., Dhingra, S., Sharma, P., Islam, S., Mohammed, I., Naina Mohamed, I., Hanna, P. A., Hajj, S. el, Jamaluddin, N. A. H., Salameh, P., & Roques, C. (2021). Surveillance of antimicrobial resistance in low- and middle-income countries: a scattered picture. In *Antimicrobial Resistance and Infection Control* (Vol. 10, Issue 1). BioMed Central Ltd. https://doi.org/10.1186/s13756-021-00931-w

Jayatilleke, K. (2020). Challenges in implementing surveillance tools of high-income countries (HICs) in low middle income countries (LMICs). *Current Treatment Options in Infectious Diseases, 12*(3), 191–201. https://doi.org/10.1007/s40506-020-00229-2

Jha, D. N. (2014, September 27). Stop prescribing antibiotics for fever and cold, Indian medical association will telll doctors. *Times of India* .

Jian, Z., Zeng, L., Xu, T., Sun, S., Yan, S., Yang, L., Huang, Y., Jia, J., & Dou, T. (2021). Antibiotic resistance genes in bacteria: Occurrence, spread, and control. *Journal of Basic Microbiology, 61*(12), 1049–1070. https://doi.org/https://doi.org/10.1002/jobm.202100201

Johnson, A. P. (2015). Surveillance of antibiotic resistance. *Philosophical Transactions of the Royal Society B: Biological Sciences, 370*(1670). https://doi.org/10.1098/rstb.2014.0080

KKRI. (2017). *National Action Plan on Antimicrobial Resistance Indonesia 2017-2019.* https://cdn.who.int/media/docs/default-source/antimicrobial-resistance/amr-spc-npm/nap-library/national-action-plan-on-antimicrobial-resistance-indonesia-2017-2019.pdf?sfvrsn=a0533d94_1&download=true

Kumar, K., Gupta, S. C., Baidoo, S. K., Chander, Y., & Rosen, C. J. (2005). Antibiotic uptake by plants from soil fertilized with animal manure. *Journal of Environmental Quality, 34*(6), 2082–2085. https://doi.org/https://doi.org/10.2134/jeq2005.0026

Kumar, S. G., Adithan, C., Harish, B. N., Sujatha, S., Roy, G., & Malini, A. (2013). Antimicrobial resistance in India: A review. *Journal of Natural Science, Biology, and Medicine, 4*(2), 286.

Kumar, S. G., Tripathi, V. R., & Garg, S. K. (2013). Antibiotic resistance and genetic diversity in water-borne Enterobacteriaceae isolates from recreational and drinking water sources. *International Journal of Environmental Science and Technology, 10*(4), 789–798. https://doi.org/10.1007/s13762-012-0126-7

Kümmerer, K. (2004). Resistance in the environment. *Journal of Antimicrobial Chemotherapy, 54*(2), 311–320. https://doi.org/https://doi.org/10.1093/jac/dkh325

Kümmerer, K. (2009a). Antibiotics in the aquatic environment – A review – Part I. *Chemosphere, 75*(4), 417–434. https://doi.org/https://doi.org/10.1016/j.chemosphere.2008.11.086

Kümmerer, K. (2009b). Antibiotics in the aquatic environment – A review – Part II. *Chemosphere, 75*(4), 435–441. https://doi.org/https://doi.org/10.1016/j.chemosphere.2008.12.006

Lange, C., Abubakar, I., Alffenaar, J. W. C., Bothamley, G., Caminero, J. A., Carvalho, A. C. C., Chang, K. C., Codecasa, L., Correia, A., Crudu, V., Davies, P., Dedicoat, M., Drobniewski, F., Duarte, R., Ehlers, C., Erkens, C., Goletti, D., Günther, G., Ibraim, E., … Cirillo, D. M. (2014). Management of patients with multidrugresistant/ extensively drug-resistant tuberculosis in Europe: A TBNET consensus statement. *European Respiratory Journal, 44*(1), 23–63. https://doi.org/10.1183/09031936.00188313

Larsson, D. G. J., de Pedro, C., & Paxeus, N. (2007). Effluent from drug manufactures contains extremely high levels of pharmaceuticals. *Journal of Hazardous Materials, 148*(3), 751–755. https://doi.org/https://doi.org/10.1016/j.jhazmat.2007.07.008

Larsson, D. G. J., & Flach, C. F. (2022). Antibiotic resistance in the environment. In *Nature Reviews Microbiology* (Vol. 20, Issue 5, pp. 257–269). Nature Research. https://doi.org/10.1038/s41579-021-00649-x

Lata, P., Ram, S., & Shanker, R. (2016). Multiplex PCR based genotypic characterization of pathogenic vancomycin resistant Enterococcus faecalis recovered from an Indian river along a city landscape. *SpringerPlus, 5*(1). https://doi.org/10.1186/s40064-016-2870-5

Laxminarayan, R., & Chaudhury, R. R. (2016). Antibiotic resistance in India: Drivers and opportunities for action. *PLoS Medicine, 13*(3). https://doi.org/10.1371/journal.pmed.1001974

Laxminarayan, R., Matsoso, P., Pant, S., Brower, C., Røttingen, J.-A., Klugman, K., & Davies, S. (2016). Access to effective antimicrobials: A worldwide challenge. *The Lancet, 387*(10014), 168–175. https://doi.org/https://doi.org/10.1016/S0140-6736(15)00474-2

Li, B., Qiu, Y., Zhang, J., Liang, P., & Huang, X. (2019). Conjugative potential of antibiotic resistance plasmids to activated sludge bacteria from wastewater treatment plants. *International Biodeterioration & Biodegradation*, *138*, 33–40. https://doi.org/https://doi.org/10.1016/j.ibiod.2018.12.013

Liguori, K., Keenum, I., Davis, B. C., Calarco, J., Milligan, E., Harwood, V. J., & Pruden, A. (2022). Antimicrobial Resistance monitoring of water environments: A framework for standardized methods and quality control. *Environmental Science and Technology*, *56*(13), 9149–9160. https://doi.org/10.1021/acs.est.1c08918

Liu, Z., Deng, D., Lu, H., Sun, J., Lv, L., Li, S., Peng, G., Ma, X., Li, J., Li, Z., Rong, T., & Wang, G. (2020). Evaluation of machine learning models for predicting antimicrobial resistance of actinobacillus pleuropneumoniae from whole genome sequences. *Frontiers in Microbiology*, *11*. https://doi.org/10.3389/fmicb.2020.00048

Lübbert, C., Baars, C., Dayakar, A., Lippmann, N., Rodloff, A. C., Kinzig, M., & Sörgel, F. (2017). Environmental pollution with antimicrobial agents from bulk drug manufacturing industries in Hyderabad, South India, is associated with dissemination of extended-spectrum beta-lactamase and carbapenemase-producing pathogens. *Infection*, *45*(4), 479–491. https://doi.org/10.1007/s15010-017-1007-2

Lundborg, C. S., & Tamhankar, A. J. (2017). Antibiotic residues in the environment of South East Asia. *BMJ (Online)*, *358*, 42–45. https://doi.org/10.1136/bmj.j2440

Maillard, J.-Y. (2018). Resistance of bacteria to biocides. *Microbiology Spectrum*, *6*(2). https://doi.org/10.1128/microbiolspec.arba-0006-2017

Martinez, J. L. (2009). The role of natural environments in the evolution of resistance traits in pathogenic bacteria. In *Proceedings of the Royal Society B: Biological Sciences* (Vol. 276, Issue 1667, pp. 2521–2530). Royal Society. https://doi.org/10.1098/rspb.2009.0320

Mason, D. J., Stott, I., Ashenden, S., Weinstein, Z. B., Karakoc, I., Meral, S., Kuru, N., Bender, A., & Cokol, M. (2017). Prediction of antibiotic interactions using descriptors derived from molecular structure. *Journal of Medicinal Chemistry*, *60*(9), 3902–3912. https://doi.org/10.1021/acs.jmedchem.7b00204

Medlicott, K., Wester, A., Gordon, B., Montgomery, M., Tayler, E., Sutherland, D., Schmoll, O., Souza, M. de, Koo-Oshima, S., DaBalogh, K., Ferreira, J. P., Erlacher-Vindel, E., Clayton, H., Graham, D., Larsson, D. J., Medema, G., Husman, A. M. R. de, Schmitt, H., Yang, M., & Zhang, Y. (2020). *Technical Brief on Water, Sanitation, Hygiene and Wastewater Management to Prevent Infections and Reduce the Spread of Antimicrobial Resistance.* https://www.fao.org/3/ca9120en/ca9120en.pdf

Meng, M., Li, Y., & Yao, H. (2022). Plasmid-mediated transfer of antibiotic Resistance genes in soil. In *Antibiotics* (Vol. 11, Issue 4). MDPI. https://doi.org/10.3390/antibiotics11040525

Merlin, C. (2020). Reducing the consumption of antibiotics: Would that be enough to slow down the dissemination of resistances in the downstream environment? *Frontiers in Microbiology*, *11*. https://doi.org/10.3389/fmicb.2020.00033

Mobarki, N., Almerabi, B., & Hattan, A. (2019). Antibiotic Resistance crisis. *International Journal of Medicine in Developing Countries*, 561–564. https://doi.org/10.24911/ijmdc.51-1549060699

MoHFW, G. (2017a). *National Health Policy, 2017.* https://cdsco.gov.in/opencms/opencms/system/modules/CDSCO.WEB/elements/download_file_division.jsp?num_id=MTU1MA==

MoHFW, G. (2017b). *National Action Plan on Antimicrobial Resistance.* https://ncdc.gov.in/WriteReadData/l892s/File645.pdf

Morgan, D. J., Okeke, I. N., Laxminarayan, R., Perencevich, E. N., & Weisenberg, S. (2011). Non-prescription antimicrobial use worldwide: A systematic review. *The Lancet Infectious Diseases*, *11*(9), 692–701. https://doi.org/https://doi.org/10.1016/S1473-3099(11)70054-8

Mutiyar, P. K., & Mittal, A. K. (2014). Risk assessment of antibiotic residues in different water matrices in India: Key issues and challenges. *Environmental Science and Pollution Research*, *21*(12), 7723–7736. https://doi.org/10.1007/s11356-014-2702-5

Nava Lara, R. A., Aguilera-Mendoza, L., Brizuela, C. A., Peña, A., & del Rio, G. (2019). Heterologous machine learning for the identification of antimicrobial activity in human-targeted drugs. *Molecules*, *24*(7). https://doi.org/10.3390/molecules24071258

Naylor, N. R., Atun, R., Zhu, N., Kulasabanathan, K., Silva, S., Chatterjee, A., Knight, G. M., & Robotham, J. v. (2018). Estimating the burden of antimicrobial resistance: A systematic literature review. *Antimicrobial Resistance and Infection Control*, *7*, 58. https://doi.org/10.1186/s13756-018-0336-y

Naylor, N. R., Zhu, N., Hulscher, M., Holmes, A., Ahmad, R., & Robotham, J. v. (2017). Is antimicrobial stewardship cost-effective? A narrative review of the evidence. In *Clinical Microbiology and Infection* (Vol. 23, Issue 11, pp. 806–811). Elsevier B.V. https://doi.org/10.1016/j.cmi.2017.06.011

Nguyen, M., Brettin, T., Long, S. W., Musser, J. M., Olsen, R. J., Olson, R., Shukla, M., Stevens, R. L., Xia, F., Yoo, H., & Davis, J. J. (2018). Developing an in silico minimum inhibitory concentration panel test for Klebsiella pneumonia. *Scientific Reports*, *8*(1). https://doi.org/10.1038/s41598-017-18972-w

Nguyen, M., Wesley Long, S., McDermott, P. F., Olsen, R. J., Olson, R., Stevens, R. L., Tyson, G. H., Zhao, S., & Davisa, J. J. (2019). Using machine learning to predict antimicrobial MICs and associated genomic features for nontyphoidal Salmonella. *Journal of Clinical Microbiology*, *57*(2). https://doi.org/10.1128/JCM.01260-18

Nkengasong, J. N., Nsubuga, P., Nwanyanwu, O., Gershy-Damet, G. M., Roscigno, G., Bulterys, M., Schoub, B., DeCock, K. M., & Birx, D. (2010). Laboratory systems and services are critical in global health: Time to end the neglect? In *American Journal of Clinical Pathology* (Vol. 134, Issue 3, pp. 368–373). https://doi.org/10.1309/AJCPMPSINQ9BRMU6

Nordmann, P., Poirel, L., Walsh, T. R., & Livermore, D. M. (2011). The emerging NDM carbapenemases. *Trends in Microbiology*, *19*(12), 588–595. https://doi.org/10.1016/j.tim.2011.09.005

Okeke, I. N., Klugman, K. P., Bhutta, Z. A., Duse, A. G., Jenkins, P., O'Brien, T. F., Pablos-Mendez, A., & Laxminarayan, R. (2005). Antimicrobial resistance in developing countries. Part II: Strategies for containment. *The Lancet Infectious Diseases*, *5*(9), 568–580. https://doi.org/10.1016/S1473-3099(05)70217-6

Ombelet, S., Ronat, J.-B., Walsh, T., Yansouni, C. P., Cox, J., Vlieghe, E., Martiny, D., Semret, M., Vandenberg, O., Jacobs, J., Lunguya, O., Phoba, M.-F., Lompo, P., Phe, T., Kariuki, S., Newton, P. N., Dance, D. A. B., Muvunyi, C., el Safi, S., … Aidara Kane, A. (2018). Clinical bacteriology in low-resource settings: today's solutions. *The Lancet Infectious Diseases*, *18*(8), e248–e258. https://doi.org/https://doi.org/10.1016/S1473-3099(18)30093-8

O'Neill, J., & Team. (2014). *Antimicrobial Resistance: Tackling a Crisis for the Health and Wealth of Nations*. https://amr-review.org/sites/default/files/AMR%20Review%20Paper%20-%20Tackling%20a%20crisis%20for%20the%20health%20and%20wealth%20of%20nations_1.pdf

O'Neill, J., & Team. (2016). *Tackling Drug-Resistant Infections Globally: Final Report And Recommendations The Review On Antimicrobial Resistance Chaired By Jim O'neill*. https://amr-review.org/sites/default/files/160518_Final%20paper_with%20cover.pdf

Opintan, J. A., Newman, M. J., Arhin, R. E., Donkor, E. S., Gyansa-Lutterodt, M., & Mills-Pappoe, W. (2015). Laboratory-based nationwide surveillance of antimicrobial resistance in Ghana. *Infection and Drug Resistance*, *8*, 379–389. https://doi.org/10.2147/IDR.S88725

Pal, C., Asiani, K., Arya, S., Rensing, C., Stekel, D. J., Larsson, D. G. J., & Hobman, J. L. (2017). Metal resistance and its association with antibiotic resistance. In R. K. Poole (Ed.), *Microbiology of Metal Ions* (Vol. 70, pp. 261–313). Academic Press. https://doi.org/https://doi.org/10.1016/bs.ampbs.2017.02.001

Pfizer. (2020). *Antimicrobial Surveillance*. Pfizer Web Page . https://www.pfizer.com/science/focus-areas/anti-infectives/antimicrobial-surveillance

Pokharel, S., & Adhikari, B. (2020). Antimicrobial resistance and over the counter use of drugs in Nepal. *Journal of Global Health*, *10*(1). https://doi.org/10.7189/JOGH.10.010360

Poonia, S., Singh, T. S., & Tsering, D. C. (2014). Antibiotic susceptibility profile of bacteria isolated from natural sources of water from rural areas of East Sikkim. *Indian Journal of Community Medicine*, *39*(3), 156–160. https://doi.org/10.4103/0970-0218.137152

Prestinaci, F., Pezzotti, P., & Pantosti, A. (2015). Antimicrobial resistance: A global multifaceted phenomenon. In *Pathogens and Global Health* (Vol. 109, Issue 7, pp. 309–318). Maney Publishing. https://doi.org/10.1179/2047773215Y.0000000030

Rabaan, A. A., Alhumaid, S., al Mutair, A., Garout, M., Abulhamayel, Y., Halwani, M. A., Alestad, J. H., Bshabshe, A. al, Sulaiman, T., Alfonaisan, M. K., Almusawi, T., Albayat, H., Alsaeed, M., Alfaresi, M., Alotaibi, S., Alhashem, Y. N., Temsah, M. H., Ali, U., & Ahmed, N. (2022). Application of artificial intelligence in combating high antimicrobial resistance rates. In *Antibiotics* (Vol. 11, Issue 6). MDPI. https://doi.org/10.3390/antibiotics11060784

Ragan, M. A., & Beiko, R. G. (2009). Lateral genetic transfer: Open issues. *Philosophical Transactions of the Royal Society B: Biological Sciences*, *364*(1527), 2241–2251. https://doi.org/10.1098/rstb.2009.0031

Rather, T. A., Hussain, S. A., Bhat, S. A., Shah, S. N., Arshid, S., & Shahnawaz, M. (2013). Antibiotic sensitivity of E. coli and Salmonella isolated from different water sources in Kashmir, India. *Comparative Clinical Pathology*, *22*(4), 729–731. https://doi.org/10.1007/s00580-012-1474-x

Reddy, B., & Dubey, S. K. (2019). River Ganges water as reservoir of microbes with antibiotic and metal ion resistance genes: High throughput metagenomic approach. *Environmental Pollution*, *246*, 443–451. https://doi.org/https://doi.org/10.1016/j.envpol.2018.12.022

Rehman, M. S. U., Rashid, N., Ashfaq, M., Saif, A., Ahmad, N., & Han, J.-I. (2015). Global risk of pharmaceutical contamination from highly populated developing countries. *Chemosphere, 138*, 1045–1055. https://doi.org/https://doi.org/10.1016/j.chemosphere.2013.02.036

Reynolds, C. A., Finkelstein, J. A., Ray, G. T., Moore, M. R., & Huang, S. S. (2014). Attributable healthcare utilization and cost of pneumonia due to drug-resistant streptococcus pneumonia: A cost analysis. *Antimicrobial Resistance and Infection Control, 3*(1). https://doi.org/10.1186/2047-2994-3-16

Rezaei-hachesu, P., Samad-Soltani, T., Yaghoubi, S., GhaziSaeedi, M., Mirnia, K., Masoumi-Asl, H., & Safdari, R. (2018). The design and evaluation of an antimicrobial resistance surveillance system for neonatal intensive care units in Iran. *International Journal of Medical Informatics, 115*, 24–34. https://doi.org/10.1016/j.ijmedinf.2018.04.007

Samreen, Ahmad, I., Malak, H. A., & Abulreesh, H. H. (2021). Environmental antimicrobial resistance and its drivers: A potential threat to public health. In *Journal of Global Antimicrobial Resistance* (Vol. 27, pp. 101–111). Elsevier Ltd. https://doi.org/10.1016/j.jgar.2021.08.001

Santajit, S., & Indrawattana, N. (2016). Mechanisms of antimicrobial resistance in ESKAPE pathogens. In *BioMed Research International* (Vol. 2016). Hindawi Limited. https://doi.org/10.1155/2016/2475067

Schuetz, P., Amin, D. N., & Greenwald, J. L. (2012). Role of procalcitonin in managing adult patients with respiratory tract infections. *Chest, 141*(4), 1063–1073. https://doi.org/https://doi.org/10.1378/chest.11-2430

Semret, M., & Haraoui, L.-P. (2019). Antimicrobial resistance in the tropics. *Infectious Disease Clinics of North America, 33*(1), 231–245. https://doi.org/https://doi.org/10.1016/j.idc.2018.10.009

Senta, I., Terzic, S., & Ahel, M. (2013). Occurrence and fate of dissolved and particulate antimicrobials in municipal wastewater treatment. *Water Research, 47*(2), 705–714. https://doi.org/https://doi.org/10.1016/j.watres.2012.10.041

Shintani, M., Nour, E., Elsayed, T., Blau, K., Wall, I., Jechalke, S., Spröer, C., Bunk, B., Overmann, J., & Smalla, K. (2020). Plant species-dependent increased abundance and diversity of IncP-1 plasmids in the rhizosphere: New insights into their role and ecology. *Frontiers in Microbiology, 11*. https://doi.org/10.3389/fmicb.2020.590776

Skandalis, N., Maeusli, M., Papafotis, D., Miller, S., Lee, B., Theologidis, I., & Luna, B. (2021). Environmental spread of antibiotic resistance. *Antibiotics, 10*(6). https://doi.org/10.3390/antibiotics10060640

Skariyachan, S., Mahajanakatti, A. B., Grandhi, N. J., Prasanna, A., Sen, B., Sharma, N., Vasist, K. S., & Narayanappa, R. (2015). Environmental monitoring of bacterial contamination and antibiotic resistance patterns of the fecal coliforms isolated from Cauvery River, a major drinking water source in Karnataka, India. *Environmental Monitoring and Assessment, 187*(5). https://doi.org/10.1007/s10661-015-4488-4

Smith, E., Lichten, C. A., Taylor, J., Maclure, C., Lepetit, L., Harte, E., Martin, A., Ghiga, I., Pitchforth, E., Sussex, J., Dujso, E., Europe, R., & Littmann, J. (2016). *Evaluation of the EC Action Plan against the Rising Threats from Antimicrobial Resistance Final Report.* https://health.ec.europa.eu/system/files/2020-01/amr_final-report_2016_rand_0.pdf

Sohail, M., Rashid, A., Aslam, B., Waseem, M., Shahid, M., Akram, M., Khurshid, M., & Rasool, M. H. (2016). Antimicrobial susceptibility of acinetobacter clinical isolates and emerging antibiogram trends for nosocomial infection management. *Revista Da Sociedade Brasileira de Medicina Tropical, 49*(3), 300–304. https://doi.org/10.1590/0037-8682-0111-2016

Sridhar, S., Turbett, S. E., Harris, J. B., & LaRocque, R. C. (2021). Antimicrobial-resistant bacteria in international travelers. In *Current Opinion in Infectious Diseases* (Vol. 34, Issue 5, pp. 423–431). NLM (Medline). https://doi.org/10.1097/QCO.0000000000000751

Swaminathan, S., Prasad, J., Dhariwal, A. C., Guleria, R., Misra, M. C., Malhotra, R., Mathur, P., Walia, K., Gupta, S., Sharma, A., Ohri, V., Jain, S., Gupta, N., Laserson, K., Malpiedi, P., Velayudhan, A., Park, B., & Srikantiah, P. (2017). Strengthening infection prevention and control and systematic surveillance of healthcare associated infections in India. *BMJ (Online), 358*, 59–62. https://doi.org/10.1136/bmj.j3768

Swapna K.M, Rajesh, R., & Lakshmanan, P. T. (2012). Incidence of antibiotic residues in farmed shrimps from the southern states of India. *Indian Journal of Geo-Marine Sciences, 41*(4), 344–347.

Szymańska, U., Wiergowski, M., Sołtyszewski, I., Kuzemko, J., Wiergowska, G., & Woźniak, M. K. (2019). Presence of antibiotics in the aquatic environment in Europe and their analytical monitoring: Recent trends and perspectives. *Microchemical Journal, 147*, 729–740. https://doi.org/https://doi.org/10.1016/j.microc.2019.04.003

Talebi Bezmin Abadi, A., Rizvanov, A. A., Haertlé, T., & Blatt, N. L. (2019). World health organization report: Current crisis of antibiotic resistance. In *BioNanoScience* (Vol. 9, Issue 4, pp. 778–788). Springer New York LLC. https://doi.org/10.1007/s12668-019-00658-4

Tamhankar, A. J., & Lundborg, C. S. (2019). Antimicrobials and antimicrobial resistance in the environment and its remediation: A global one health perspective. In *International Journal of Environmental Research and Public Health* (Vol. 16, Issue 23). MDPI AG. https://doi.org/10.3390/ijerph16234614

Taneja, N., & Sharma, M. (2019). Antimicrobial resistance in the environment: The Indian scenario. In *Indian Journal of Medical Research* (Vol. 149, Issue 2, pp. 119–128). Wolters Kluwer Medknow Publications. https://doi.org/10.4103/ijmr.IJMR_331_18

Tucker, T. J., & Manyike, P. T. (2017). Improving the clinic-laboratory-interface in the context of HIV diagnosis, treatment, and monitoring. *Current Opinion in HIV and AIDS*, *12*(2). https://journals.lww.com/co-hivandaids/Fulltext/2017/03000/Improving_the_clinic_laboratory_interface_in_the.2.aspx

Uchil, R. R., Kohli, G. S., Katekhaye, V. M., & Swami, O. C. (2014). Strategies to combat antimicrobial resistance. *Journal of Clinical and Diagnostic Research*, *8*(7). https://doi.org/10.7860/JCDR/2014/8925.4529

UNEP. (2017). *Antimicrobial Resistance: Investigating the Environmental Dimension*. https://wedocs.unep.org/bitstream/handle/20.500.11822/22263/Frontiers_2017_CH1_EN.pdf?sequence=1&isAllowed=y

UNEP. (2022). *Antimicrobial Resistance: A Global Threat*. https://www.unep.org/explore-topics/chemicals-waste/what-we-do/emerging-issues/antimicrobial-resistance-global-threat

van Boeckel, T. P., Brower, C., Gilbert, M., Grenfell, B. T., Levin, S. A., Robinson, T. P., Teillant, A., & Laxminarayan, R. (2015). Global trends in antimicrobial use in food animals. *Proceedings of the National Academy of Sciences of the United States of America*, *112*(18), 5649–5654. https://doi.org/10.1073/pnas.1503141112

Ventola, C. L. (2015). The antibiotic resistance crisis: part 1: causes and threats. P & T: A peer-reviewed journal for formulary management, 40(4), 277–283.

Voermans, A. M., Mewes, J. C., Broyles, M. R., & Steuten, L. M. G. (2019). Cost-effectiveness analysis of a procalcitonin-guided decision algorithm for antibiotic stewardship using real-world U.S. Hospital data. *OMICS A Journal of Integrative Biology*, *23*(10), 508–515. https://doi.org/10.1089/omi.2019.0113

Weinstein, Z. B., Bender, A., & Cokol, M. (2017). Prediction of synergistic drug combinations. In *Current Opinion in Systems Biology* (Vol. 4, pp. 24–28). Elsevier Ltd. https://doi.org/10.1016/j.coisb.2017.05.005

Wellington, E. M. H., Boxall, A. B. A., Cross, P., Feil, E. J., Gaze, W. H., Hawkey, P. M., Johnson-Rollings, A. S., Jones, D. L., Lee, N. M., Otten, W., Thomas, C. M., & Williams, A. P. (2013). The role of the natural environment in the emergence of antibiotic resistance in Gram-negative bacteria. *The Lancet Infectious Diseases*, *13*(2), 155–165. https://doi.org/10.1016/S1473-3099(12)70317-1

WHO. (2011). *Jaipur Declaration on Antimicrobial Resistance*. https://apps.who.int/iris/bitstream/handle/10665/205397/B4828.pdf?sequence=1&isAllowed=y

WHO. (2014a). *Antimicrobial Resistance: Global Report on Surveillance*. https://www.who.int/publications/i/item/9789241564748

WHO. (2014b). *Global Tuberculosis Report 2014* (Vol. 1). https://apps.who.int/iris/bitstream/handle/10665/137094/9789241564809_eng.pdf?sequence=1&isAllowed=y

WHO. (2021, November 17). *Antimicrobial Resistance* . https://www.who.int/news-room/fact-sheets/detail/antimicrobial-resistance

WHO. (2022). *Global Tuberculosis reporT 2022*. http://apps.who.int/bookorders.

WHO-ROSEA. (2011). *Jaipur Declaration on Antimicrobial Resistance*. https://apps.who.int/iris/bitstream/handle/10665/205397/B4828.pdf?sequence=1&isAllowed=y

WHO-ROSEA. (2017). *Situational Analysis on Antimicrobial Resistance in the South-East Asia Region*. https://apps.who.int/iris/bitstream/handle/10665/272873/AMR-2016report.pdf?sequence=1&isAllowed=y

Wright, G. D. (2007). The antibiotic resistome: the nexus of chemical and genetic diversity. *Nature Reviews Microbiology*, *5*(3), 175–186. https://doi.org/10.1038/nrmicro1614

Yuan, Q. bin, Guo, M. T., & Yang, J. (2015). Fate of antibiotic resistant bacteria and genes during wastewater chlorination: Implication for antibiotic resistance control. *PLoS One*, *10*(3). https://doi.org/10.1371/journal.pone.0119403

Zhang, N., Juneau, P., Huang, R., He, Z., Sun, B., Zhou, J., & Liang, Y. (2021). Coexistence between antibiotic resistance genes and metal resistance genes in manure-fertilized soils. *Geoderma*, *382*, 114760. https://doi.org/https://doi.org/10.1016/j.geoderma.2020.114760

8 Antimicrobial Resistance in the Environment and Their Impact

N. Sathammai Priya, R. Thenmozhi,
P. F. Steffi, and B. Thirumalaiyammal

8.1 INTRODUCTION

Numerous bacterial species evolved the capability to tolerate antibiotics long before humans started to mass-produce them to help treat contagious conditions. Insulated caves, permafrost cores, and other environments and samples that have been saved from anthropogenic bacterial contamination can give insight into the resistance mechanisms that prevailed during the pre-antibiotic period. An important driver of the ancient and still ongoing evolution of resistance mechanisms is likely to be the never-ending competition for resources among microorganisms, including the natural production of secondary metabolites that are analogous to many of the antibiotics used today as medicinals. The fairly recent introduction of antibiotics as clinical agents radically changed the conditions for the evolution and spread of resistance by providing unprecedented selection pressures, especially on members of the microbiota of humans and domestic animals, but also in environments heavily contaminated with antibiotics. This selection pressure has promoted the accumulation and vertical transfer of a wide range of antibiotic resistance genes (ARGs) to numerous bacterial species, particularly those causing diseases. The ultimate, well-known consequences of such accumulating evolutionary events are gradually increasing difficulties in preventing and treating bacterial infections. As bacteria and genes frequently cross environments and species boundaries, it is critical to understand and recognize the connections between human, animal, and environmental microbiota (the "One Health" concept) to manage this global health challenge (WHO 2020).

Typically focuses on AMR in pathogens but limited perspective. Across all territories, the total number of bacterial species alone may exceed 1 million and only a tiny bit of these, some 10–20 species such as *Mycobacterium tuberculosis* or Staphylococcus pneumoniae, are specialist mortal pathogens. A slightly larger number, several hundred species, are opportunistic pathogens that only cause human disease in certain circumstances, such as *Listeria* spp., *Campylobacter* spp. or *Staphylococcus aureus*. Hundreds more are part of the normal human microbiota, the majority of which are commensals that have not been linked to disease. Many of the pathogenic bacteria (and presumably many of the commensals too) are not only in humans but also in other hosts, a wide variety of animal and wildlife species (that is, they're zoonotic) or in the wider environment (sapronotic) or both, *Escherichia coli* being an obvious example (Neill J 2014).

8.2 PROVENANCE OF RESISTANCE

8.2.1 COMMENSALS

The arrival of metagenomics has provided insight into the makeup of human gut foliage. One study found numerous resistance genes in the unculturable portion, the so-called microbial "dark matter" that comprises the bulk of the gut flora, but these were not homologous with resistance genes of

 DOI: 10.1201/9781003268925-8

clinical applicability. By contrast, the culturable portion contained numerous homologues of resistance genes in pathogenic bacteria. However, the direction of transfer between the commensal gut flora and the pathogens is uncertain.

8.2.2 Soil

The global biomass of microbes is enormous. A crude computation harmonious with published estimates gives a value of roughly 50 tonnes of bacteria per person. Most of this biomass is found in soil, and soil is also the original source of the majority of antibiotics used in human and veterinary medicine. Soil bacteria, and other soil microbes, have been producing antibiotics on a global scale for maybe 2 billion years. Soil, the most biodiverse niche on Earth, contains the most diverse microorganisms, and within that diversity harbours ARGs or their precursors, which contribute to the emergence and proliferation of antibiotic resistance. Moreover, human activity exposes the soil to various pollutants, facilitating the selection of existing genes, resistant mutants or horizontal gene transfer (HGT) between soil microbiota. Antibiotic resistance has become the prominent illustration of the significance of the "One Health" perspective, including the three main disciplines: human health, the environment, and animal husbandry. Soil, as a critical core of the "One Health" framework, not only contains a massive natural resistome but also receives numerous ARGs and antibiotic-resistant bacteria (ARB) from animal and mortal wastes, which could be cycled back to human-dominated surroundings.

8.2.2.1 Livestock Sector

Antibiotics are essential veterinary drugs used for animals to treat bacterial infections, and to help control disease in a herd and at sub-therapeutic concentrations for growth promotion. They are administered either orally via feed or throughdrinking water, or parenterally. Antibiotics are ranked in significance differently for human medicine and veterinary medicine. In the many cases where the same antibiotic class is critically important to both sectors, e.g. fluoroquinolones and third- and fourth-generation cephalosporins, recommendations for veterinary use are provided by the World Organisation for Animal Health (OIE), namely a) they shouldn't be used for metaphylaxis (preventing specific disease outbreaks); b) they should not to be used as a first-line treatment unless justified and should be guided by antimicrobial susceptibility testing; c) off-label use should be limited and reserved for when no alternatives are available; and d) they should not to be used as growth promotors (Taneja and Sharma, 2019).

The impact of AMR on livestock health has received far lower attention than that on human health. Analogous to humans, AMR in creatures can lead to suffering from infections that would not have otherwise occurred, an increased frequency of treatment failures, and increased severity of infections. Likewise, losing treatment options, either through the development of resistance or through restrictions on their use, will have consequences for livestock health and welfare. For possessors of food and commodity-producing creatures, AMR may lead to financial losses directly through higher mortality and indirectly through reduced production and growth, decreased feed conversion, as well as early culling of breeding and product creatures. ultimately this may lead to higher prices of goods and food from livestock products for the end consumer (ICMR 2020).

8.2.2.2 Environmental Transmission

The environs can provide a route for some resistant bacteria to populate or infect hosts. The operation of ordure in husbandry, which introduces large quantities of ARB into soil, significantly changes the structure of the soil microbial community. The transfer of ARGs through associated mobile inheritable elements, such as transposons, plasmids, and genomic islets, to other microbial populations increases the frequency of ARB in the environment. Elimination of ordure-born microbiota could effectively mitigate the proliferation of ARGs in manured soil, and different indigenous soil microorganisms may help prevent the irruption and colonization of ordure-born ARB in soil.

The majority of ARB was temporarily settled in microbiota, whereas a minority became permanently settled. Temporary colonization might lead to the transfer of ARGs to clinical pathogens. For illustration, soil-inhabiting bacteria showed a high nucleotide identity to those in clinical mortal pathogens(Purohit M. R. et al., 2019).

Food and contamination on surfaces, especially in hospitals, are two ways in which certain resistant bacteria spread. A variety of infections can also be contracted from exposure to surface waters contaminated by faecal residues. The European Union, for instance, uses levels of faecal indicator bacteria as a surrogate exposure threshold in the bathing water directive. There is no reason to believe that resistant strains will not be exposed to the same exposure risk as sensitive strains through contaminated water. Consequently, correlative analyses suggest that the sanitation and waste infrastructure plays a greater role in determining national levels of resistance than antibiotic use.

However, only a few studies have identified the potential role of environmental exposure as a factor in colonizing and/or infecting specific resistant bacteria. Taking faecal swabs from surfers in the United Kingdom, who ingest seawater more often than non-surfers, revealed a greater likelihood of cephalosporin-resistant *E. Coli.* Several studies have found that recreational swimming could increase the risk of urinary tract infections caused by extended-spectrum β-lactamase-producing *E. coli* or K*lebsiella pneumoniae.* The extent to which contaminated water plays a role in the transmission of resistant bacteria needs further research in both high- and low-income settings (Singh et al., 2014).

With data on environmental emissions, fate, and exposure, it is possible to quantify the risks for the transmission of resistant bacteria. The environmental route is still poorly described in accurate models. Models based on specific resistant bacteria are likely to be more predictive than models based on ARGs, since the bacterial host and genetic context of mobile ARGs are critical indicators of transmission risk. The dosing requirements for colonization by many bacteria remain an important knowledge gap.

Environmental transmission is unclear, if at all, when selective agents play a role. The environment can be a hostile environment for some pathogens (such as *Legionella* or *Vibrio*) even though some thrive in it (such as *Legionella*). Environmental factors often limit the growth of those pathogens. Therefore, it is conceivable that small differences in growth rates between antimicrobial-resistant strains and non-resistant strains, caused by sub-MICs of antibiotics, may determine whether an environment becomes sufficiently high for infection or colonization in humans or animals. Besides antibiotic resistance profiles, other environmental transmission opportunities for both resistant and non-resistant strains are likely to be much more important due to factors such as temperature, oxygen pressure, nutrients, predation, and competition with other species (Marti et al., 2014).

8.3 ENVIRONMENTAL PATHWAYS OF ANTIMICROBIAL RESISTANCE

The emergence and spread of AMR in the environment are facilitated by a complex web of pathways through which resistant bacteria and resistance genes are disseminated. Understanding these pathways is crucial for devising effective strategies to mitigate the environmental impact of AMR. This chapter examines the key environmental pathways that contribute to the dissemination of AMR, highlighting their interconnectedness and implications (Berendonk T. U. et al .,2015).

8.3.1 Wastewater Discharge

Wastewater treatment plants receive effluents containing antibiotics, antibiotic-resistant bacteria, and resistance genes from various sources, including households, hospitals, and industries. While treatment processes can reduce bacterial loads, they may not eliminate all resistant bacteria and genes, leading to their release into aquatic ecosystems through treated effluents (Baquero F. et al., 2008).

8.3.2 Agricultural Runoff

The application of manure and fertilizers from animals treated with antibiotics can lead to the introduction of resistant bacteria and genes into soil and water. Rainfall and irrigation can transport these contaminants into nearby water bodies, contributing to the spread of resistance. Agricultural fields can serve as reservoirs where resistant bacteria interact with indigenous microbes.

8.3.3 Aquaculture and Mariculture Effluents

Antibiotics used in aquaculture to prevent and treat diseases can be released into aquatic environments. Resistant bacteria can emerge and thrive in these settings, and their effluents can contribute to the dissemination of resistance to other environments, including natural water bodies.

8.3.4 Pharmaceutical Manufacturing Discharges

Pharmaceutical production facilities release antibiotic residues, intermediate compounds, and resistant bacteria into wastewater. These discharges can contribute to the enrichment of environmental reservoirs with resistant bacteria and genes. The proximity of manufacturing sites to water bodies enhances the potential for AMR dissemination.

8.3.5 Horizontal Gene Transfer

Environmental settings promote the transfer of resistance genes between bacterial species through horizontal gene transfer mechanisms such as conjugation, transformation, and transduction. Mobile genetic elements like plasmids and transposons facilitate the transfer of resistance traits across diverse bacterial populations, amplifying the spread of AMR.

8.3.6 Wildlife and Animal Interactions

Wildlife, including birds and rodents, can carry resistant bacteria and genes from human-impacted environments to natural ecosystems. These animals can act as vectors, introducing AMR to remote areas. Additionally, interactions between domestic animals, wildlife, and humans can lead to the exchange of resistant strains (ICMR 2020).

8.3.7 International Travel and Trade

Global travel and trade contribute to the widespread dissemination of AMR. Resistant bacteria can be carried by travellers, animals, and imported goods, introducing novel resistance strains to different geographic locations. This interconnectedness underscores the need for coordinated global efforts to tackle environmental AMR (C.D.P. 2016).

8.3.8 Recreational Water Activities

Recreational water bodies can harbour resistant bacteria due to contamination from human and animal waste. Activities like swimming and boating can lead to direct exposure, potentially allowing the transfer of resistant bacteria from water bodies to human skin and mucous membranes. Understanding these diverse pathways of AMR dissemination is essential for designing interventions to mitigate the impact of environmental AMR. A multifaceted approach involving improved waste management, responsible antibiotic use, and cross-disciplinary collaboration is crucial to address this global challenge (Blair J. M. et al., 2015).

8.4 SOURCES OF ENVIRONMENTAL ANTIMICROBIAL RESISTANCE

8.4.1 Pathogens with Community-Acquired Resistance

8.4.1.1 *S. pneumoniae*

There are about a million deaths attributed to pneumococci among children under the age of 5 in India every year. The penicillin-binding protein (PBP) is altered in pneumococcal bacteria to make them resistant to beta-lactams. A downward revision was made to the breakpoints for defining meningeal and non-meningeal resistance to penicillin and cephalosporin in 2008, whereas an upward revision was made for non-meningeal resistance.

8.4.1.2 *S. aureus*

S. aureus caused in children by community-acquired (CA) gets into the skin, soft tissue, bones, and joints, and causes pneumonia. Through PBP alteration, CA pathogens are becoming increasingly resistant to methicillin. An investigation from Pondicherry found that 20% of children with pyoderma had MRSA. In one study, children from Central India were colonized with MRSA in their nasal passages in 29%. Children with serious skin and soft tissue infections, bone and joint infections, and necrotizing pneumonia should be treated with anti-MRSA drugs empirically due to the rising prevalence of CA-MRSA.

8.4.1.3 *Shigella*

Among children younger than 5 years old living in developing countries, particularly in India, Shigella is responsible for 5–10% of diarrhoea cases. India also has high prevalence of fluoroquinolone resistance, apart from cotrimoxazole and ampicillin resistance. As a result, cefixime and azithromycin are the most common antimicrobials prescribed in India for bloody diarrhoea. Infection of other enteric pathogens by plasmid-borne cephalosporin resistance is a concern due to the development of cephalosporin resistance (Appelbaum P. C. 1992).

8.5 SOLUTIONS TO THE PROBLEM OF ANTIMICROBIAL RESISTANCE

As a result of antimicrobial resistance, preventing infections (which will reduce antibiotic use indirectly) is key to solving the problem. Antimicrobial resistance can also be tracked (currently, ICMR does this effectively), rational antimicrobial therapy can be promoted, and new drugs must be developed.

8.5.1 Prevent Infections

AMR will be reduced by this strategy, as well as childhood morbidity and mortality. There are a number of interventions in this area, including improving immunization services, providing safe water, improving sanitation, reducing pollution, promoting hand hygiene, breastfeeding, and improving nutrition. A priority is also being placed on typhoid vaccine introduction in the national programme in addition to pneumococcal and rotavirus vaccines being introduced incrementally (Laxminarayan R. et al., 2016).

8.5.2 Rational Use of Antimicrobials

There are many stakeholders involved, including the government, pharmaceutical companies, medical professionals, the media, and the public (Figure 8.1). Legislation controlling the use of antibiotics in animal agriculture is the role of the government and regulatory agencies. A link between medical licence renewal and antimicrobial therapy training will help practising physicians upgrade their knowledge. Along with rationalizing antimicrobial therapy, making diagnostics more affordable and available at point of care will facilitate accurate diagnosis. Medical guidelines must be

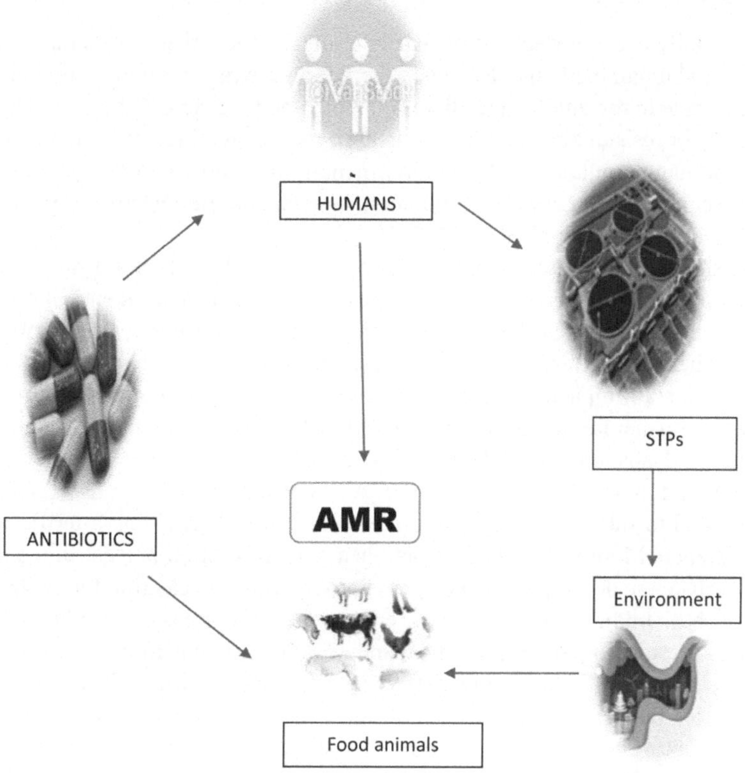

FIGURE 8.1 Provenance of antimicrobial resistance.

developed and disseminated to the community so that infections can be managed effectively (ICMR recently published such a guideline). Despite the foregoing discussion, many gaps still exist in our understanding of AMR's relationship with the environment. Nevertheless, there is adequate evidence to support the conclusion that there are several environmental health risks for humans and animals and that coordinated efforts are needed to address them. In order to ensure proper AMR surveillance, several sources have identified the lack of quantitative data as a key barrier (Rinke C. et al., 2013).

In spite of the complexity of assessing environmental risks, interagency and interjurisdictional risk assessments can aid in identifying risks and management goals, along with clear indicators of efficiency to develop environmental management plans. A recent study suggested that environmental antibiotic resistance must be assessed in addition to traditional methods. According to the study, antibiotic-resistant bacteria must be considered and accounted for by taking into account the selective pressures added to the environmental resistome over time; identifying and describing the extent of HGT in the relevant environmental hot spots; and modifying dose-response approaches to evaluate the impact of antibiotic resistance on certain health outcomes and pathways. It is necessary to conduct risk assessment studies to ensure that wastewater effluents do not pose health or environmental hazards due to their high concentration of resistant bacteria. To accurately assess the level of resistant bacteria in wastewater effluents that could pose a health risk to humans and the environment, risk assessment studies are required. It is also necessary to develop wastewater treatment technologies that produce effluent with bacteria at an acceptable level of resistance. It is possible to estimate the changes in environmental ecosystems in clinical settings through routine sampling of organic soils and residues in hospital drains, and to enhance existing antimicrobial stewardship efforts (Morar and Wright, 2010) (Figure 8.2).

8.6 ENVIRONMENTAL DIMENSION OF ANTIMICROBIAL RESISTANCE

AMR has traditionally been viewed as a primarily clinical issue, arising from the overuse and misuse of antibiotics in human medicine. However, there is a growing recognition that the environment plays a significant role in the emergence, dissemination, and persistence of AMR. The environment acts as a reservoir for resistant bacteria and resistance genes, creating opportunities for their transfer to and from human populations. This environmental dimension of AMR is shaped by complex interactions between natural ecosystems, human activities, and the selective pressures imposed by anthropogenic factors (Klumper U. et al. 2015).

Studies have shown that environmental bacteria, including those found in soil, water, and wildlife, can harbour a diverse range of resistance genes, some of which are identical or closely related to those found in clinical settings. These genes can confer resistance to a variety of antibiotics and can be exchanged between different bacterial species through horizontal gene transfer mechanisms. The proximity of human-impacted environments, such as agricultural areas and urban centres, to natural ecosystems further facilitates the spread of AMR. The widespread use of antibiotics in agriculture, for instance, leads to the introduction of antimicrobial agents into the environment through irrigation, runoff, and the application of manure. As a consequence, both susceptible and resistant bacteria are exposed to sub-lethal concentrations of antibiotics, providing a fertile ground for the selection of resistance. Moreover, pollutants and heavy metals, often present in the same environments, can exacerbate the development and spread of AMR by co-selecting for antibiotic resistance traits. In light of these findings, addressing the environmental dimension of AMR is paramount. A comprehensive understanding of the ecological and evolutionary dynamics of resistance emergence and transmission is essential for developing effective strategies to mitigate its impact. Implementing regulatory measures that restrict the release of antibiotics and other pollutants into the environment, as well as adopting sustainable agricultural and waste management practices, are crucial steps (Heuer and Smalla, 2012).

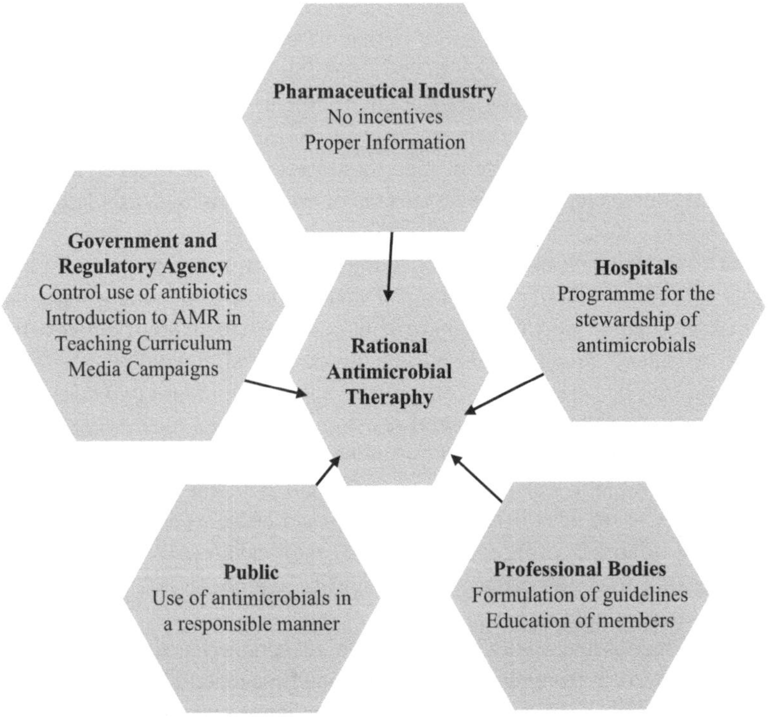

FIGURE 8.2 Rational use of antibacterials.

8.7 ENVIRONMENTAL FACTORS ENHANCING ANTIMICROBIAL RESISTANCE

AMR is a complex issue influenced by various environmental factors that contribute to the selection, proliferation, and dissemination of resistant bacteria and resistance genes. These factors interact in intricate ways, creating conditions that foster the development and spread of AMR within environmental microbial communities (Schijven J. F. et al., 2015).

8.7.1 Pollution and Chemical Stressors

Pollution, including the presence of heavy metals, industrial chemicals, and pesticides, can exert selective pressure on bacteria. Some of these pollutants might share similar resistance mechanisms with antibiotics, leading to cross-resistance. Additionally, pollutants can disrupt bacterial cell membranes, facilitating the uptake of resistance genes from the environment or other bacteria. This interaction between pollutants and antibiotic resistance highlights the importance of addressing both types of stressors in combating AMR (O'Neill J. 2014).

8.7.2 Biocides and Disinfectants

The widespread use of biocides and disinfectants, intended to control microbial populations, can inadvertently promote AMR. These agents can act as drivers of selection for bacteria with intrinsic or acquired resistance mechanisms. Bacteria that survive biocide exposure might also display increased tolerance to antibiotics. The constant exposure to biocides in settings like hospitals and households underscores the potential link between biocide use and AMR.

8.7.3 Horizontal Gene Transfer

Environmental conditions, such as nutrient availability and stress, can enhance horizontal gene transfer – the process by which resistance genes are transferred between bacteria. Increased stress, often caused by anthropogenic activities, can trigger genetic responses that promote the transfer of resistance genes. Consequently, environments rich in organic matter or those affected by pollution can serve as hotspots for gene transfer events, accelerating the spread of AMR (Sethi S. et al., 2006).

8.7.4 Antibiotic Residues and Selective Pressure

The presence of antibiotic residues in the environment, whether from agricultural runoff or pharmaceutical manufacturing effluents, exerts a strong selective pressure on bacterial populations. Sublethal concentrations of antibiotics can lead to the survival and proliferation of resistant bacteria. This phenomenon is especially concerning in environments with inadequate waste treatment systems, as it amplifies the exposure and potential for resistance development.

8.7.5 Climate Change and Dispersal Patterns

Climate change can influence bacterial behaviour and survival, affecting their potential for AMR development. Altered temperatures and precipitation patterns can impact bacterial growth rates, survival times, and dispersal patterns. These changes might lead to shifts in bacterial communities, potentially favouring the proliferation of resistant strains in specific environmental niches.

In light of these intricate relationships, effective management of environmental factors enhancing AMR requires a multifaceted approach. Strategies include improving waste treatment processes to reduce the release of antibiotics and pollutants, regulating the use of biocides, and considering

AMR risks when designing pollution control measures. By addressing the environmental drivers of AMR, we can complement clinical efforts and contribute to the comprehensive fight against antimicrobial resistance.

8.8 IMPACT OF AGRICULTURAL PRACTICES ON ENVIRONMENTAL AMR

Agricultural practices play a significant role in shaping the emergence and dissemination of AMR in the environment. The widespread use of antibiotics in livestock, aquaculture, and crop production has led to the introduction of antimicrobial agents into ecosystems, creating conditions conducive to the development and spread of resistance (Pal C. et al., 2016).

8.8.1 Antibiotic Use in Agriculture

The routine administration of antibiotics to livestock for growth promotion and disease prevention is a common practice in modern agriculture. This practice leads to the release of antibiotic residues and resistant bacteria into the environment through manure and waste runoff. Consequently, agricultural settings become reservoirs for resistant bacteria, which can potentially spread to humans through direct contact or contaminated food.

8.8.2 Selection Pressure and Co-Selection

Antibiotic use in agriculture exerts selective pressure on microbial populations. This pressure favours the survival and proliferation of bacteria with resistance traits, contributing to the development of AMR. Furthermore, the presence of pollutants and heavy metals in agricultural environments can create co-selection scenarios, where resistance to multiple agents is promoted simultaneously.

8.8.3 Aquaculture and Aquatic Environments

Antibiotic use in aquaculture, particularly in densely stocked fish farms, has become a significant concern. The release of antibiotic-laden water from these farms can lead to the enrichment of nearby aquatic environments with both antibiotics and resistant bacteria. These environments provide opportunities for horizontal gene transfer and the propagation of resistance within aquatic microbial communities.

8.8.4 Horizontal Gene Transfer in Agricultural Soils

Agricultural soils can act as reservoirs for antibiotic-resistant bacteria and resistance genes. Horizontal gene transfer mechanisms, such as conjugation and transformation, enable the spread of resistance traits among soil bacteria. The use of manure from antibiotic-treated animals as fertilizer can introduce resistant bacteria and genes into soils, perpetuating the cycle of resistance.

8.8.5 The "One Health" Approach

The interconnectedness of human health, animal health, and environmental health is emphasized through the "One Health" approach. Addressing agricultural AMR necessitates collaboration among human health professionals, veterinarians, agronomists, and environmental scientists. Implementing responsible antibiotic use in agriculture, promoting animal welfare, and adopting agroecological practices can collectively contribute to reducing the environmental impact of AMR (Perry J. et al., 2016).

8.8.6 Regulatory and Educational Interventions

Stricter regulations on antibiotic use in agriculture, coupled with educational programmes for farmers and stakeholders, can promote awareness of the consequences of AMR. Encouraging the adoption of alternative disease prevention strategies, such as vaccines and probiotics, can help reduce the reliance on antibiotics in agriculture.

As the global demand for food continues to grow, it becomes imperative to balance agricultural productivity with the preservation of environmental and human health. By acknowledging and addressing the link between agricultural practices and environmental AMR, we can work towards a sustainable future where agricultural systems contribute to, rather than undermine, the fight against antimicrobial resistance (Gaze W. H. et al., 2013).

8.9 ECOLOGICAL AND PUBLIC HEALTH CONSEQUENCES OF ENVIRONMENTAL AMR

The growing threat of AMR in the environment carries profound ecological and public health implications. As resistant bacteria and resistance genes spread through ecosystems, they disrupt ecological balances and pose risks to human and animal health (Pruden A. et al., 2017).

8.9.1 Disruption of Microbial Communities

AMR can alter the composition and diversity of microbial communities in various environments. Resistant bacteria might outcompete susceptible counterparts, leading to shifts in ecosystem dynamics. These changes can impact nutrient cycling, organic matter decomposition, and other crucial ecological processes, potentially leading to cascading effects throughout the ecosystem.

8.9.2 Impacts on Aquatic Systems

Aquatic environments are particularly vulnerable to the consequences of AMR due to the direct release of antibiotics and resistant bacteria. Aquatic organisms can be exposed to antibiotics through water and sediments, potentially leading to the development of resistance. Disruption of aquatic microbial communities can impact water quality, nutrient cycling, and the health of aquatic organisms.

8.9.3 Zoonotic and Foodborne Transmission

AMR in the environment increases the likelihood of zoonotic transmission, where resistant bacteria and genes are transferred from animals to humans. Consumption of contaminated food and water can introduce resistant pathogens into the human body. This transmission can result in difficult-to-treat infections and limit the efficacy of antibiotics.

8.9.4 Healthcare Challenges and Treatment Failures

Environmental AMR complicates the treatment of infections in clinical settings. Resistant bacteria from the environment can find their way into hospitals and healthcare facilities, leading to healthcare-associated infections. These infections are difficult to treat and can lead to prolonged hospital stays, increased healthcare costs, and even mortality.

8.9.5 Dissemination of Resistance

Environmental reservoirs of resistant bacteria and genes serve as sources for the spread of AMR within and between ecosystems. Horizontal gene transfer can lead to the rapid dissemination of resistance, making containment efforts challenging. Resistant bacteria originating from the environment can colonize humans, animals, and other environments, perpetuating the cycle of resistance.

8.9.6 Economic and Societal Impact

The ecological and health consequences of environmental AMR can have far-reaching economic and societal impacts. Increased healthcare costs, loss of productivity, and reduced agricultural yields are just a few examples. The potential for treatment failures and increased mortality further underscore the urgency of addressing environmental AMR.

8.9.7 Conservation and Ecosystem Services

AMR-induced disruptions in ecosystems can impact the provision of essential ecosystem services. For instance, changes in microbial communities can affect soil fertility, water purification, and disease regulation. These changes can have implications for agriculture, clean water availability, and overall ecosystem resilience.

The multifaceted nature of the consequences of environmental AMR underscores the importance of adopting a comprehensive approach to address this issue. Combating AMR requires collaborative efforts across disciplines, industries, and countries. By recognizing the interconnectedness of ecological and public health impacts, society can work towards sustainable solutions that safeguard both human health and the integrity of the environment (Pruden A. et al., 2017).

8.10 AMR IN INDIA

India has faced a growing problem of antibiotic resistance due to various factors:

1. **High antibiotic consumption:** India is one of the largest consumers of antibiotics globally, with both appropriate and inappropriate use in healthcare settings contributing to the development of antibiotic-resistant bacteria.
2. **Over-the-counter availability:** Antibiotics are often available without a prescription, leading to misuse and overuse by individuals who may not fully understand their proper usage.
3. **Limited surveillance:** Surveillance of antibiotic resistance and bacterial infections is not uniform across the country, making it challenging to track and manage the issue effectively. Inadequate infection prevention and control measures in healthcare facilities can lead to the spread of resistant bacteria among patients.
4. **Lack of public awareness:** Limited awareness among the general public about antibiotic resistance and responsible antibiotic use has contributed to the problem.

8.10.1 Environmental Implications

Antibiotic resistance is not confined to human health; it also has implications for the environment.

1. **Wastewater contamination:** Antibiotics, antibiotic-resistant bacteria, and antibiotic residues from human waste, agricultural runoff, and pharmaceutical manufacturing can contaminate wastewater, leading to the spread of resistance genes in the environment.

2. **Agricultural use:** The use of antibiotics in agriculture can contribute to the emergence of antibiotic-resistant bacteria in soil and water. Antibiotics are sometimes used in animal husbandry for growth promotion and disease prevention.
3. **Aquatic ecosystems:** Antibiotic-resistant bacteria and residues can enter aquatic ecosystems through wastewater discharge and runoff, potentially affecting aquatic organisms and causing the spread of resistance.
4. **Food safety concerns:** Antibiotic-resistant bacteria can be present in food products originating from animals treated with antibiotics, posing a risk to human health when consumed.

8.10.2 Efforts and Solutions

Addressing antibiotic resistance in India requires a multifaceted approach:

1. **Regulation:** Strengthening regulations on antibiotic use in healthcare and agriculture, including prescription-only access, and monitoring antibiotic consumption.
2. **Antibiotic stewardship:** Promoting responsible antibiotic use through educational campaigns targeting healthcare professionals, patients, and the public.
3. **Infection control:** Improving infection prevention and control practices in healthcare settings to reduce the spread of resistant bacteria.
4. **Surveillance:** Enhancing surveillance systems to monitor antibiotic resistance trends and track emerging resistant strains.
5. **Wastewater treatment:** Implementing advanced wastewater treatment methods to reduce the release of antibiotics and resistant bacteria into the environment.
6. **Research:** Conducting research to understand the mechanisms and extent of antibiotic resistance in different settings and environments.
7. **Global collaboration:** Collaborating with international organizations and other countries to share knowledge, experiences, and best practices in combating antibiotic resistance.

8.11 CONCLUSION

There are several parallels between the challenge of tackling AMR and other twenty-first-century crises, such as climate change, as recently pointed out. In the last half-century or so, human activity has been able to influence both AMR and climate change on a global scale. Although there are some similarities, there are also a number of differences. There are already available and being used on a significant scale alternatives to burning fossil fuels in the context of climate change. Health economics, international law, and social science are also clearly needed to add input to the research, in addition to epidemiology, microbiology, and pharmacology. Nevertheless, governments, industries, international agencies, as well as scientists, will need to coordinate in order to take effective action against AMR. In order to achieve a long-term solution, clinicians, pharmacists, patients, veterinarians, and farmers will need to participate in the process, as all of them have contributed to the current AMR problem. A key priority is developing strategies for better antimicrobial stewardship (both in human medicine and industry), reinvigorating the drug pipeline for antimicrobials, and developing effective and sustainable alternatives to combat infectious diseases in both humans and livestock.

REFERENCES

Appelbaum, P. C. (1992). Antimicrobial resistance in *Streptococcus pneumoniae*: An overview. *Clinical Infectious Diseases*, 15, 77–83.

Baquero, F., & Martinez, J. L. (2008). Antibiotics and antibiotic resistance in water environments. *Current Opinion in Biotechnology*, 19(3), 260–265.

Berendonk, T. U., Manaia, C. M., Merlin, C., Fatta-Kassinos, D., Cytryn, E., Walsh, F., ... Schwartz, T. (2015). Tackling antibiotic resistance: The environmental framework. *Nature Reviews Microbiology*, 13(5), 310–317.

Blair, J. M., Webber, M. A., Baylay, A. J., Ogbolu, D. O., & Piddock, L. J. (2015). Molecular mechanisms of antibiotic resistance. *Nature Reviews Microbiology*, 13(1), 42–51.

Center for Disease Dynamics Economics and Policy. Resistance Map.(http://resistancemap.cddep.org/resmap/c/in/India, accessed 20 June, 2016).1. Center for Disease Dynamics Economics and Policy. Resistance Map.(http://resistancemap.cddep.org/resmap/c/in/India, accessed 20 June, 2016).

Gaze, W. H., Krone, S. M., & Larsson, D. G. (2013). Pollution: An overlooked pathway for the release of antibiotics from livestock. *Environmental Science & Technology*, 47(18), 9600–9601.

Heuer, H., & Smalla, K. (2012). An overview of the influence of antibiotics and antibiotic resistance genes on the bacterial community in the environment. *Environmental International*, *49*, 3-15. https://doi.org/10.1016/j.envint.2012.08.007

Indian Council for Medical Research. (2020). *Antimicrobial Resistance Research and Surveillance Network*. Indian Council for Medical Research.

Jain, S., & Chugh, T. D. (2013, November 15). Antimicrobial resistance among blood culture isolates of Salmonella enterica in New Delhi. *Journal of Infection in Developing Countries*, 7(11), 788–795.

Klumper, U., et al. (2015). Broad host range plasmids can invade an unexpectedly diverse fraction of a soil bacterial community. *The ISME Journal*, 9, 934–945.

Laxminarayan, R., Chaudhury, R. R., Laxminarayan, R., Duse, A., Wattal, C., Zaidi, A., et al. (2016, March 2). Antibiotic resistance in India: Drivers and opportunities for action. *PLoS Medicine*, 13(3), e1001974.

Marti, E., Variatza, E., & Balcázar, J. L. (2014). The role of aquatic ecosystems as reservoirs of antibiotic resistance. *Trends in Microbiology*, 22(1), 36–41.

Morar, M., & Wright, G. D. (2010). Environmental antibiotic resistance: A comprehensive review. *Environmental Health Perspectives*, 118(8), 1082-1091. https://doi.org/10.1289/ehp.0901900

O'Neill, J. (2014). *Review on antimicrobial resistance. Antimicrobial resistance: Tackling a crisis for the health and wealth of nations*. London: Review on Antimicrobial Resistance.

Pal, C., Bengtsson-Palme, J., Kristiansson, E., & Larsson, D. G. J. (2016). The structure and diversity of human, animal and environmental resistomes. *Microbiome,* 4, 54. doi:10.1186/s40168-016-0199-5.

Perry, J., Waglechner, N., & Wright, G. (2016). The prehistory of antibiotic resistance. *Cold Spring Harbor Perspectives in Medicine*, 6, a025197.

Pruden, A., Larsson, D. G., & Amézquita, A. (2017). Antibiotic resistance genes as emerging contaminants: Studies in Northern Colorado. *Environmental Science & Technology*, 51(9), 4939–4948.

Purohit, M. R., Lindahl, L. F., Diwan, V., Marrone, G., & Lundborg, C. S. (2019). High levels of drug resistance in commensal E. coli in a cohort of children from rural central India. *Scientific Reports*, 9, 6682.

Rinke, C., et al. (2013). Insights into the phylogeny and coding potential of microbial dark matter. *Nature*, 499, 431–437.

Schijven, J. F., Blaak, H., Schets, F. M., & De Roda Husman, A. M. (2015). Fate of extended-spectrum β-lactamase-producing Escherichia coli from faecal sources in surface water and probability of human exposure through swimming. *Environmental Science & Technology*, 49, 11825–11833.

Sethi, S., Sharma, D., Mehta, S. D., Singh, B., Smriti, M., Kumar, B., et al. (2006, May). Emergence of ciprofloxacin resistant Neisseria gonorrhoeae in north India. *Indian Journal of Medical Research*, 123(5), 707–710.

Singh, P., Kaur, R., & Kaur, R. (2014). Antimicrobial Resistance in the Environment: The Indian Scenario. *Indian Journal of Microbiology*, 54(3), 291-302.

Taneja, N., & Sharma, S. (2019). Antibiotic use in veterinary medicine: A review. *Veterinary World*, *12*(1), 14-20. https://doi.org/10.14202/vetworld.2019.14-20

Wellington, E. M., Boxall, A. B., Cross, P., Feil, E. J., Gaze, W. H., Hawkey, P. M., ... Williams, A. P. (2013). The role of the natural environment in the emergence of antibiotic resistance in gram-negative bacteria. *The Lancet Infectious Diseases*, 13(2), 155–165.

World Health Organization. (2020). *Antimicrobial resistance*. Geneva: World Health Organization.

9 Antimicrobial Resistance and Multiple Drug Resistance in Microorganisms

Source, Evolution, and Mitigation Approaches for Preventing Health Hazards

Ragini Arora, Bikash Kumar, and Harit Jha

9.1 INTRODUCTION

Antibiotics have been widely employed in disease control and cattle production in recent decades. Due to the widespread usage of antibiotics, there is an ongoing issue of antibiotic residue discharge into the environment (Zhang et al., 2009). The presence and dissemination of antibiotics in higher concentrations in the environment through different sources (wastewater, medical disposal, etc.) have led to antibiotic resistance in bacteria. Antibiotic-resistant bacteria (ARB) includes microorganisms resistant to several drugs and carrying one or more antibiotic-resistant genes (ARGs) that can emerge as a result of antibiotic overuse and improper disposal. ARBs and ARGs are increasingly recognized as emerging contaminants that endanger food safety and human health (Naquin et al., 2015). The World Health Organization has highlighted antibiotic resistance as a major public health issue. A post-antibiotic future is conceivable, as per the World Health Organization (WHO), if the issue is not given significant consideration o (WHO, 2023). Antimicrobial-resistant illnesses may cause more fatalities by 2050 than any other key factor. Opportunistic infections in individuals receiving medical attention for any other pathology have made antibiotic resistance a severe hazard to human health (Karkman et al., 2016).

Antibiotic resistance, on the other hand, is not just a healthcare concern because it spreads among domestic and farm animals, healthy people, wildlife, and natural habitats, particularly water and soil (McEwen & Collignon, 2018; Hernando-Amado et al., 2019). The consequences of widespread diffusion and inevitable interaction with ARB are heavily influenced by the complex context in which they occur, ranging from socioeconomic and climatic circumstances to other elements that are yet little understood (Vaz-moreira et al., 2022). Even in industrialized nations, common infectious illnesses that are today curable with medications may soon become fatal and incurable. If costly therapies are required for common illnesses, the situation will be considerably worse in Third World emerging nations. ARGs found in peripatetic genetic components of pathogenic bacteria are particularly concerning since they pose a major risk of developing into incurable diseases and spreading further (Karkman et al., 2016). The spread of multi-drug resistance (MDR) in populations of both non-pathogenic and pathogenic bacteria is an interrelated problem with major implications for community health. Higher rates of morbidity and mortality, higher treatment costs, limited

DOI: 10.1201/9781003268925-9

treatment options, longer hospital stays, and a greater need for novel antibacterial agents are just a few of the public health concerns brought on by the development of AMR and MDR in pathogenic bacteria (Mukherjee et al., 2021).

Antibiotics have been frequently observed to be present in the environment, particularly in groundwater, surface water, and sediment (Grehs et al., 2021). One of the primary sources of antibiotic-resistant bacteria and the release of resistance genes into the environment is wastewater. It serves as an environmental provider of antibiotic resistance and acts as reservoir and hotspot for horizontal gene transfer (HGT), allowing antibiotic resistance genes to spread even further (Rizzo et al., 2013) In wastewaters, an increase in horizontal gene transfer is observed because of high bacterial populations and generated biofilms. Also, stress produced by pollutant chemicals such as heavy metals and antibiotics forces HGT for the natural survival of the bacterial community. Municipal wastewater turns into an essential breeding ground for resistant organisms or genes since 30–90% of all antimicrobials are excreted unaltered through human faeces and urine. Municipal wastewater that may be "AMR-rich" is released into the neighbouring bodies of water.

Gram-negative bacteria were found in these north Indian rivers, and resistance genes including blaNDM-1 and blaOXA48 were found in them. Contamination of water bodies happens as a result of improper antibiotic disposal, which is increasing due to population growth, urbanization, and industry. The antibiotic contamination in the environment is mainly caused by the release of incompletely metabolized antibiotics by humans and animals; additional contributors include waste from pharmaceutical production operations and the disposal of unwanted medications (Kumar & Pal, 2017).

The presence of different carbon sources and other nutrients in wastewater, along with the presence of oxygen or nitrate, and particles that can be adsorbed by bacteria, a pH of 6.6–7.4, and stable temperatures (15–20°C) are crucial for bacterial survival. High bacterial densities and the potential for a variety of medications to be present in wastewater can make it easy for bacteria to develop antibiotic resistance and release responsible ARGs into the surrounding environment. These released ARGs can easily spread to environmental bacteria and diseases through genetic change, raising the dangers and hazards to humans, animals, and the humans, animals and environment (Liu et al., 2012). Additionally, the treatment of wastewater has been shown to be largely ineffective against specific *Enterococci* strains that are resistant to erythromycin, ciprofloxacin, and tetracycline. Nevertheless, the treatment process has been shown to increase the prevalence of ciprofloxacin resistance (da Silva et al., 2006).

9.2 CAUSES OF ANTIBIOTIC RESISTANCE

As the globe becomes increasingly industrialized and urbanized, the amount of antibiotics in the environment increases. The environment becomes progressively contaminated as a result of human activities, including mining, garbage disposal, industrialization, and others. Industrial effluents are now being released into the environment at a higher rate as a result of population expansion and industrial development. Heavy metals and antibiotics found in wastewater have the potential to cause heavy metal and antibiotic resistance. Public health is at risk from ARB and heavy metal-resistant microorganisms and their related genes. Emerging pollutants such as ARGs, heavy metals, antibiotics, and heavy metal resistance genes (HMRGs) are commonly found in municipal wastewater. According to research, reducing antibiotic use does not always reduce the transmission and maintenance of AR since factors other than antibiotics can increase cross-resistance. For example, bacteria have evolved metal tolerance mechanisms such as (1) metal complexation or sequestration, (2) intracellular ion reduction via proteins (e.g. MerA protein reduction of Hg^{2+} to Hg^0, and (3) efflux systems. Although antibiotic resistance arises naturally and is an unavoidable result of even cautious antibiotic usage, it is known that abuse and misappropriation of antibiotics is the primary predictor of the increase in antibiotic resistance. Intense antibiotic therapy may assist the individual patient, but also has environmental and ecological repercussions, unlike any other medicine category (Seiler & Berendonk, 2012).

9.3 MECHANISMS OF ANTIBIOTIC RESISTANCE

The ability of bacteria to withstand the negative impact of antimicrobial treatments is called antimicrobial resistance. Bacteria's genetic plasticity, which causes adaptive responses, permits genetic mutations and the acquisition of genetic material through horizontal gene transfer (transformation, transduction, and conjugation), which ensures the widespread distribution of bacterial species and determines antimicrobial resistance. Different antimicrobial resistance-coding genes are found in the genomes of bacteria, reflecting genotypic resistance. Under the selection pressure of the antibiotic, the expression of AGRs results in the activation of resistance mechanisms (phenotype resistance), including active expulsion systems, biofilm formation, cell wall recycling, efflux pumps, porin reduction, and target protein alteration (Bair et al., 2020).

9.3.1 Antibiotic Modification

Basically there are two different mechanisms of bacterial defence against antibiotics: (i) bacteria release enzymes that can either inactivate the antibiotics by adding a specific chemical group or (ii) degrade the molecules, making it unavailable to interact with their target (Munita et al., 2016). Different bacterial cell wall components are usually the targets for antibiotics, and modification in these components make them resistant to the antibiotics treatment. For example, variations in the quantity and/or structure of penicillin-binding proteins (PBPs) make them resistant to β-lactam medicines. This mechanism is essentially employed by Gram-positive bacteria having penicillin-binding proteins.

PBPs are transpeptidases that assist in constructing the cell wall peptidoglycan. Changes in the number of PBPs can impact the drug binding ability to the target (i.e. a decrease in PBPs with normal drug binding or an increase in PBPs with decreased drug binding ability). Drug binding may become completely or partially inhibited by structural changes. Vancomycin is an example of a glycopeptide that works by preventing the manufacture of cell walls, while daptomycin is an example of a lipopeptide that works by depolarizing the cell membrane. These medications are intrinsically resistant to Gram-negative bacteria with extensive LPS layers (Randall CP et al., 2013).

Vancomycin resistance has grown to be a serious problem in *Staphylococcus aureus* and *Enterococci* (VRE – vancomycin-resistant *Enterococci*) (MRSA). Van genes are acquired during the development of resistance, changing the structure of peptidoglycan precursors and decreasing the capacity of vancomycin to bind to them. Calcium is necessary for the binding of daptomycin. Gene changes, such as mprF, cause the cell membrane surface to become positively charged, which prevents calcium from interacting with the membrane and, as a result, daptomycin from binding (Yang et al., 2009).

Drugs also target ribosomal subunits for their effect. Therefore, ribosomal mutation (aminoglycosides, oxazolidinones), methylation (streptogramins, oxazolidinones, aminoglycosides, macrolides – Gram-positive bacteria), or ribosomal protection (tetracyclines) are a few mechanisms that make the bacteria resistant to drugs. These processes inhibit the medication from binding to the ribosome. Amid these pathways, the degree of medication interference differs considerably. Resistance to fluoroquinolones, which target the production of nucleic acids, is caused by changes to either DNA gyrase or topoisomerase IV in Gram-negative bacteria (gyrA) and Gram-positive bacteria (grlA), respectively. Gyrase and topoisomerase undergo alterations due to these mutations, which reduce or completely remove the drug's capacity to bind to these molecules.

For medications that block metabolic processes, resistance develops through the overproduction of dihydropteroate synthase (DHPS) and dihydrofolate reductase (DHFR) enzymes (trimethoprim – DHFR, sulfonamides – DHPS) or mutations in the enzymes involved in the folate biosynthesis pathway. These medications work by competitive inhibition where the drug competes with the substrates for binding to the enzymes' active site. The active location of these enzymes is most frequently where mutations occur, and the consequent structural alterations in the enzyme prevent the natural substrate from binding while interfering with drug binding (Reygaert, 2018). As is seen in producers, pathogenic organisms frequently modify antibiotics to build up resistance to aminoglycosides. Both Gram-negative and Gram-positive bacteria have been shown to have a variety of AMEs, including a fusion enzyme with both AAC and APH functions. A thorough nomenclature has also been devised. In clinical bacteria, these genes are typically identified on mobile genetic elements (MGEs). In a variety of environmental bacteria, chromosomal factors for AMEs have also been discovered, including *Acinetobacter* and *Providencia* sp., which are likely to be the source of acquired factors seen on MGEs in pathogenic strains. Both Gram-negative and Gram-positive bacteria have CAT enzymes in addition to AMEs, and they have all been thoroughly examined. β-lactamase is the degradation/modification enzyme utilized in clinical bacterial strains. While the significance of β-lactamases in producer bacteria is currently being debated, it is well established that they are essential for β-lactam resistance in clinical Gram-negative bacteria. Instead, Gram-positive bacteria choose PBP-based defence mechanisms, which is most likely because their cell walls and envelopes are built differently from those of Gram-negative bacteria (Peterson & Kaur, 2018).

9.3.2 Drug Efflux

Antibiotic efflux and/or reduced permeability is the second main mechanism of antibiotic resistance in clinical strains. The presence of the outer membrane acts as a permeability barrier, and this is crucial for Gram-negative bacteria, and provides an innate mechanism for defence against antimicrobial drugs (e.g. vancomycin) and hydrophilic antibiotics (Nikaido, 2003). Genes for efflux pumps are encoded in the bacterial chromosome, out of which several are expressed permanently, while others are induced or overexpressed in response to certain modifications/mutations (e.g. a mutation that alters the transport channel can cause high-level resistance). The efflux pumps, such as multi-drug (MDR) efflux pumps, transport/remove a wide range of hazardous molecules from the bacterial cell.

It has been demonstrated that porin gene mutations and/or variations in their expression affect the susceptibility of Gram-negative bacteria to hydrophilic drugs. Additionally, both Gram-negative and Gram-positive bacteria possess a variety of active efflux pumps, most of which fall under one of the following five families: ATP-binding cassette (ABC), Multi-drug and Toxin Extrusion (MATE), Major Facilitator Superfamily (MFS), Resistance-Nodulation-Division (RND), and Small Multi-drug Resistance (SMR). The majority of these efflux pump families are single-component pumps and penetrate the cytoplasmic membrane with the substrates. In contrast, the RND family consists of multi-component pumps mainly found in Gram-negative bacteria, which work in synchronization with the help of periplasmic membrane fusion protein (MFP) and outer membrane protein (OMP-porin) to efflux substrates across whole cell envelope (Sun et al., 2014). In some cases, efflux family members together with other cellular elements function as multi-component pumps in Gram-negative bacteria. To extrude macrolide medications, members of the ABC and MacB families (MacAB-TolC) function as a tripartite pump. EmrB, a component of the MFS, functions as a tripartite pump (EmrAB-TolC) to extrude nalidixic acid from *E. coli*, An intrinsic resistance in Gram-positive bacteria efflux pumps is imparted in Gram-positive bacteria efflux pumps, due to their chromosomal encoding. Members of the MFS and MATE families, as well as fluoroquinolone efflux pumps, are included in these pumps. Known plasmid-borne Gram-positive efflux pumps are also present. The MFS family now constitutes the described pumps in Gram-positive bacteria. All five families of efflux pumps have been found in Gram-negative bacteria, with the RND family having the most therapeutically relevant pumps (Kourtesi et al., 2013).

9.3.3 Limiting Drug Uptake

Naturally, a variation is observed in the ability of bacteria to restrict the absorption of antimicrobial drugs. The LPS layer's structure and activities in Gram-negative bacteria block several chemicals; because of that, these bacteria have inherent resistance to specific classes of powerful antimicrobial drugs (Blair et al., 2014). Since Gram-positive bacteria lack an outer barrier, limiting access to medications is less common. The *Enterococci* have an inherent resistance to aminoglycosides because polar molecules have trouble entering the cell wall. Therefore, all medications that specifically target the cell wall, such as β-lactams and glycopeptides, are fundamentally resistant/inactive against microorganisms that lack a cell wall, like *Mycoplasma.* Since Gram-positive bacteria lack an outer barrier, limiting access to medications is less common. The *Enterococci* have an inherent resistance to aminoglycosides because polar molecules have trouble entering the cell wall. The reduction of porins by these bacteria serves as a defence against carbapenem. Resistance to imipenem and several cephalosporins has been shown in *Neisseria gonorrhoeae* and *E. aerogenes* due to mutations that modify the porin channel, resulting in resistance to β-lactams and tetracycline. The creation of a biofilm by a bacterial colony is another well-known process in the colonization of bacteria. The biofilm community of the typical flora in the gut is an example of how these biofilms can include a dominating organism or a wide community of species. In terms of pathogenic organisms, the creation of a biofilm shields the bacterium against host immune system attacks as well as antimicrobial treatments (Thiolas et al., 2004).

9.3.4 Target Modification/Bypass/Protection

Antimicrobial resistance is developed by bacteria through interference at the target location. Bacteria have developed several strategies such as (i) target protection, i.e. preventing the antibiotic from reaching its binding site, and (ii) structural changes to the target site that cause decreased affinity towards the antibiotic molecule. In bacterial chromosomes, several genetic determinants coding for proteins that mediate target protection have been discovered. MGEs carry most of the therapeutically important genes implicated in this mode of resistance to antibiotics. Fluoroquinolones (Qnr), Tetracycline (Tet[M] and Tet[O]), and fusidic acid are examples of medications influenced by this mechanism (FusB and FusC).

The Tet(M) and Tet(S) are two classic and well-studied instances of tetracycline resistance determinants for target protection mechanisms. Tet(M) in *Streptococcus* spp. and Tet(O) were discovered in *Campylobacter jejuni*, but they are now observed in a variety of bacterial species. The possible reason behind their presence in several bacterial strains is that these are controlled by plasmids and the presence of broad-range conjugative transposons. The introduction of changes at the target site of antibiotics is one of the most predominant approaches of antibiotic resistance in bacterial pathogens, impacting almost all the antimicrobial chemical families. Target modification can occur through (i) alterations at the specific sites in target encoding genes, (ii) enzymatic modifications of the binding site (e.g. methyl group addition), and/or (iii) an auxiliary or bypass through the original target.

As previously stated, irrespective of the modification, the end result remains unaffected, i.e. a decrease in antibiotic's affinity for the target site. The presence of exogenous PBP, known as PBP2a, with a characteristics transpeptidase domain that is intolerant to the action of wide range of β-lactams, confers resistance to MRSA strains towards β-lactams. Bypassing the first sensitive target is made possible by the acquisition of PBP2a; although, as it lacks transglycosylase activity, it works in conjunction with the native PBP2 transglycosylase domain to carry out cross-linking activities when β-lactams are present. Enzymatic changes or point mutations to the target are examples of further target modification in clinical strains. See examples of target point mutations for further information (Hooper 2002). The presence of a significant number of erythromycin ribosomal methylation (erm) genes that provide macrolide resistance provides the best explanation for how an enzyme alters the target. In the 23S rRNA, a particular adenine is methylated by these enzymes. The erm genes are common among both Gram-negative and Gram-positive bacteria and are found on mobile genetic elements in clinical strains (Roberts, 2008)(Figure 9.1 and Table 9.1).

9.4 WASTEWATER AS A SOURCE OF MICROORGANISMS WITH ANTIBIOTIC RESISTANCE

9.4.1 Microbial Diversity

Antibiotic resistance develops, spreads, and enters the environment in wastewater systems, which are hotspots for antibiotic resistance. They are home to numerous commensal human and animal microorganisms as well as antibiotic resistance determinants. Water shortage is an issue that affects the whole world, although it is particularly acute in specific areas like Africa, the Middle East, southern Europe, and the western United States. So, irrigation with treated wastewater is presumably a key way for antibiotics to penetrate soil ecosystems in these places. Numerous studies have shown that wastewater contains bacteria and genes that are resistant to antibiotics (Piotrowska & Popowska, 2015).

In wastewater, pollutants include chemical contaminants (such as pesticides), heavy metals, disinfectants/surfactants, and antibiotic residues from diverse sources (hospitals, pharmaceutical, and domestic sectors). Common human bacteria combined with bacteria that might live in the sewage system make up the majority of the sewage microbiota. When it comes to a particular category of *Enterobacteriaceae* or *Enterococci*, the proportion of antibiotic-resistant bacteria in wastewater might reach more than 50%. The presence of the aforementioned factors results in selection pressure and the simultaneous selection phenomenon, which in turn causes horizontal gene transfer because the bioreactor system, such as municipal wastewater treatment plants, provides extremely favourable conditions for bacterial growth (Teshome et al., 2020).

Additionally, it is believed that mobile genetic elements play a substantial role in the transmission of ARG between bacterial strains. Various studies have established that the existence of chemical pollutants, such as antibiotic residues, and their unrestrained release into the environment cause the

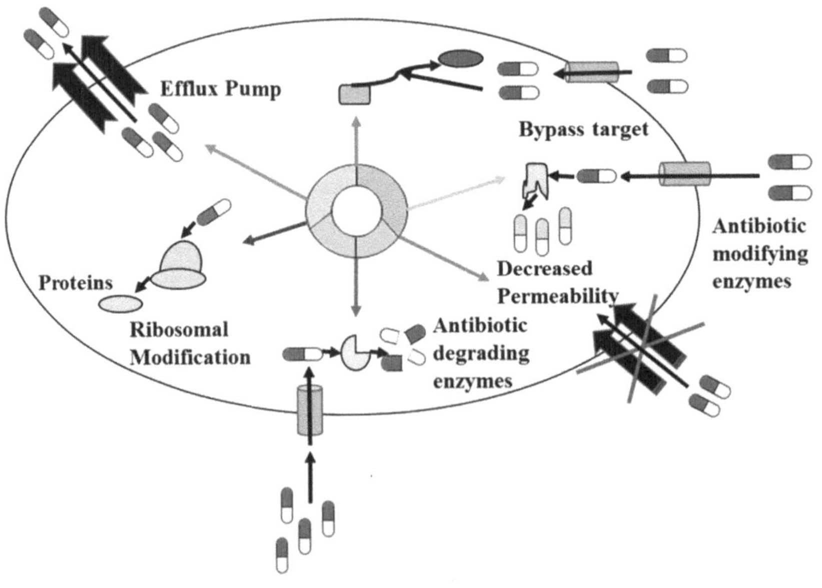

FIGURE 9.1 Pictorial representation of the mechanism followed by different microorganisms to be resistant against different antibiotics.

TABLE 9.1
Antibiotic Resistance Mechanisms of Isolates from Wastewater

Antibiotic	Host	Mechanism of Resistance	References
β-Lactams	*Pseudomonas fluorescens*	Acquired IMP	(Girlich et al., 2010)
	Acinetobacter baumannii	Acquired OXA-23	
	Enterococcus faecium	PBP5 mutations	(Schell et al., 2020)
Vancomycin	*Enterococci* spp.	Peptidoglycan modification	(Cetinkaya et al., 2000)
Chloramphenicol	*Aeromonas bestiarium*	FloR efflux	(To & Editor, 2013)
Tetracyclines	*Several species*	Tet efflux	(Grossman, 2016)
Fluoroquinolones	*E. coli*	OqxAB efflux	(Stapleton et al., 2020)
	P. aeruginosa	QRDR (quinolones resistance-determining region) mutations	(Jalal et al., 2000)
	A. punctata	QnrVC4	(Peterson & Kaur, 2018)

spread of antibiotic-resistant bacteria and associated genes, in particular antibiotic resistance genes. According to the findings of various research, wastewater is a "hot zone" for antibiotic-resistant bacteria and genes (Piotrowska et al., 2017).

Acinetobacter, *Enterococci*, and *Enterobacteriaceae*, which are on the list of the world's top priority pathogens, make up the majority of antibiotic-resistant bacteria discovered in municipal wastewater networks. Many *Enterococci* and *Enterobacteriaceae* are resistant to more than one antibiotic, according to an analysis of published studies. Additionally, it has been demonstrated that various *Escherichia coli*, *Shigella*, and *Klebsiella* bacteria increased from an average of 11% in raw wastewater to 21% in treated wastewater when they were resistant to multiple antibiotics. The percentage of *Enterobacteriaceae* that were resistant to the three medicines rose from 5.5% to 14.1% for the treated effluent (Khoshmardan & Khoshmardan, 2021).

These findings further suggest that the traditional municipal wastewater treatment method does not completely eliminate *Enterobacteriaceae* that are resistant to antibiotics. Methicillin-resistant *Staphylococcus aureus* (MRSA) was found in the effluent of four US municipal wastewater treatment facilities in 2012. Similar to activated sludge, clinically significant drugs like ciprofloxacin and vancomycin have been proven to be ineffective against bacteria (Nagulapally et al., 2009). Pure wastewater poses a harm to the public's health if it is not properly handled, as evidenced by the numerous comparable cases that are comparable (Hong et al., 2013).

Municipal wastewater sustains extremely varied microbial communities, including bacteria from the environment, human and animal commensals, and both pathogenic and non-pathogenic species. These comprise species and strains that are both vulnerable and resistant to antibiotics, as well as other types of abiotic and chemical stresses. In the past, bacterial screening of wastewater often concentrated on particular bacterial taxa, particularly medically important pathogens such as *Escherichia coli*, *Klebsiella pneumonia*, *Pseudomonas aeruginosa*, *Acinetobacter* spp., and *Aeromonas* spp. (Goni-Urriza et al., 2000; Boczek et al., 2007).

Influence on influent wastewater characteristics comes from variations in local industry, climate, population density, and demography. For instance, certain areas may be impacted by extremely high salinity brought on by industrial effluents or coastal groundwater penetration. *Actinobacteria*, *Bacteroidetes*, and *Firmicutes* follow *Proteobacteria* in terms of relative abundance in wastewater influents, according to the most recent research. These phyla contribute in varying amounts, but according to scientists, they made up around 40% of the entire influent community (McLellan et al., 2010) Although they remained the second-dominant phylum after Proteobacteria, lower proportions of *Actinobacteria*, *Bacteroidetes*, and *Firmicutes* were found in the Shatin WWTP influent (Ye & Zhang, 2013). A total of 23 phyla were identified in the effluent samples from the same plant and 19 in the influent samples; however, they highlighted that rarefaction curves suggested that further sequencing would reveal more species from these samples. *Firmicutes* made up a significant fraction (40.9%) of the influent, although this proportion decreased to 4.43% in the effluent. *Actinobacteria* and *Bacteroidetes* were more prevalent in the effluent than the influent in their investigation, but the basic conclusion that these four phyla are the most frequent was still true (Yang et al., 2014).

According to current findings, *Desulfobacterales*, *Clostridiales*, *Desulfovibrionales*, *Lactobacillales*, and *Bifidobacteriales* were the most prevalent orders in the influent sample. This was distinct from the prevalent communities of *Pseudomonas*, *Mycobacterium*, and *Vibrio* found in effluent. As with the rest of the community structure, the relative abundance of pathogens in the entire community varies in response to therapy. Several pathogen-related sequences in the effluent were dramatically decreased, whereas some pathogens, such as *Aeromonas hydrophila*, had increasing relative abundance (Voolaid et al., 2018).

9.4.2 Horizontal Gene Transfer (HGT)

One of the important forces influencing bacterial evolution is the foreign DNA acquisition by HGT, which commonly results in the emergence of antibiotic resistance. One creature obtaining the DNA of another organism is how horizontal gene transfer, also known as HGT, occurs. Many different kinds of creatures can be seen doing this, although it is often done by prokaryotic species, mostly bacteria (Munita et al., 2016). The majority of antibacterial substances utilized in clinical settings are those that naturally occur in the environment. As was previously indicated, bacteria that coexist with these chemicals have inherent genetic resistance factors. Studies show that "environmental resistomes" are a crucial source for the spread of ARG in clinically important bacteria. Additionally, genetic exchange has been proven to be an evident source for the spread of antibiotic resistance. The "endosymbiotic" creature having both sets of DNA is produced by this process, which can be carried out in a variety of ways. Bacteria carry out HGT primarily through three different mechanisms: transformation, conjugation, and phage-mediated transduction. Although each HGT mechanism is often employed in nature, some bacterial species favour one mechanism over the others. For instance, *Enterobacteria* frequently employ conjugative plasmids for genetic information exchange, but streptococci can become spontaneously competent and so efficiently engage in transformation (Peterson & Kaur, 2018).

9.4.2.1 Transformation

The simplest form of HGT is probably transformation, although only a few clinically significant bacterial species have the capacity to "naturally" assimilate bare DNA to produce resistance (Munita et al., 2016) Natural transformation is the key method of integrating DNA from different organisms. Bacteria allow a piece of their genetic material (i.e. DNA) escape and it is later picked up by a different, nearby bacterium. This process is being watched. It may then be incorporated into the DNA of the organism, which can be exploited to develop new capabilities.

This procedure is necessary for cell speciation, adaptability, and cellular evolution, just like it is for other HGT techniques. It can help transfer DNA to another organism, increasing pathogenicity and enhancing metabolic response efficiency. In particular, this process has the potential to introduce a significant, novel defence mechanism: antibiotic resistance. Although many Gram-negative bacteria also transform, Gram-positive bacteria like *Streptococcus pneumoniae* and *Bacillus subtilis* are better known for it. In fact, transformation appears to have been a key factor in the development of *Streptococcus* and *Neisseria* strains resistant to antibiotics. For instance, it is believed that the high frequency of spontaneous transformation in *S. pneumoniae* may be responsible for the maintenance of penicillin resistance in this bacterium (Hoffman Roberts et al., 2005).

9.4.2.2 Conjugation

Conjugation is an effective gene transfer method that includes cell-to-cell contact. It occurs at very high rates in the gastrointestinal tract of individuals undergoing antibiotic therapy. This method is often involved in the development of ARG in the hospital setting (Graf et al., 2019). Mobile genetic elements are frequently able to transfer important genetic information during conjugation; however, direct transfer from one chromosome to another is well documented. Transposons and plasmids are important MGEs because they contribute significantly to the emergence and spread of antibiotic resistance in therapeutically relevant organisms.

This kind of HGT happens when two bacterial cells come into intimate contact. During conjugation, separation of the DNA takes place via the action of bi-functional enzymes with inherent trans-esterase activity. The DNA is then moved through the covalent link between the two cells and unwound by the action of DNA helicase enzymes before entering the new bacterium and becoming incorporated into its own DNA composition. However, compared to either transformation or transduction, plasmid-mediated conjugation is still thought to be the most common in nature for spreading resistance genes. Plasmids may replicate independently and comprise genetic factors for resistance to several important antibiotics. In actuality, transposons that are carried by plasmids consist of resistance gene collections, providing resistance to several antibiotic classes and metal ions at the same time. Moreover, based on the host range of the plasmid, they have the capacity to transfer genes to several species, genera, and even kingdoms across vast genetic distances (Peterson & Kaur, 2018).

9.4.2.3 Phage-Mediated Transduction

Viruses that infect bacterial cells are known as "phages", or bacteriophages. A subclass of phages, i.e. "temperate bacteriophages", are capable of transferring the DNA of one bacterium into another. This process is called transduction mediated by phages. Sometimes these phase-mediated transduction involves the transfer of genes responsible for antibiotic resistance and virulence. Three different environmental bacteriophages were identified and infected *Enterococcus faecalis* and *Enterococcus gallinarum* bacterial species. These bacteria, along with phages, were then allowed to grow together, providing suitable environment for transduction to occur. These bacteria were then subjected to various antimicrobial chemicals after transduction, and it was found that the majority of *Enterococcus faecalis* and *Enterococcus gallinarum* bacteria developed tetracycline and gentamicin resistance, respectively. In light of this, it was shown that phage-mediated transduction significantly contributes to bacterial colonies' development of antibiotic resistance. Although it has been demonstrated that many bacteria experience transduction at a modest frequency of between 106 and 109 transductants/plaque-forming-unit, transduction is believed to have had a substantial impact on *S. aureus*' development of resistance (Ubukata et al., 1975). It is commonly known that transduction in particular and HGT in general play a substantial role in the spread of antibiotic resistance genes in *S. aureus*, the accessory genome of the highly strain-variable *S. aureus* contains many transposons, plasmids, phages, SCCmec, and genomic islands (Haaber et al., 2017) (Figure 9.2).

The primary methods by which bacterial species can mobilize and exchange genetic material with both related and unrelated species are as outlined above. These processes suggest a method for bacterial evolution in various settings, enabling them to persist in their niches. The evolution of bacteria with ability to have inherent mechanisms against antibiotics and the ability to possess virulence to survive and grow in newly evolved or harsh environments.

9.4.3 Evolution of Antibiotic Resistance in Microbial Consortium

Ability to resist or survive under unfavourable conditions or exposure to materials that have antibiotic properties has most certainly existed since the existence of bacteria's ubiquitous nature (probably

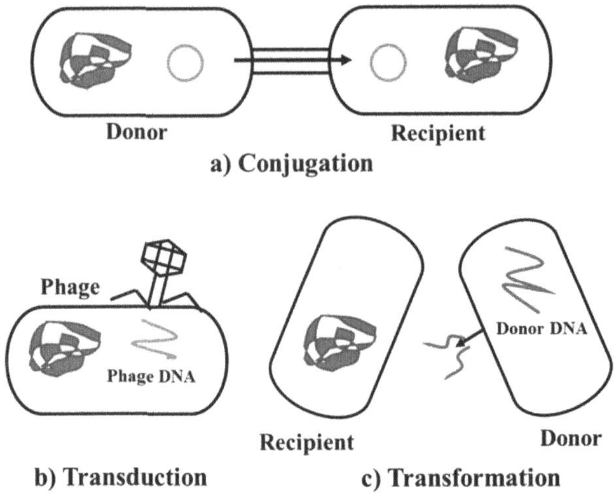

FIGURE 9.2 This figure elaborates on three mechanisms followed for horizontal gene transfer where figures a, b and c represent conjugation, transduction, and transformation, respectively.

3 billion years). While we humans tend to contemplate these antibiotics as life-saving medications/drugs, in reality these drugs are usually derived from naturally occurring substances produced by microbes. The microbes produce these biochemicals naturally for their own "selfish" purposes, i.e. in order to fend off other bacteria or to survive and grow in the presence of other microbes. The natural application of these biochemicals is entirely separate from bacterial inhibition. In reaction to the existence of these antibiotics, highly specific resistance genes have developed (Vikesland et al., 2017) Antibiotic Resistance in bacteria, as well as how it spreads and appears, is a global phenomenon. A deeper comprehension of the emergence of motives and hot spots involved in the evolution and dissemination of novel antibiotic resistances or multi-drug resistances is immediately required. Darwinian factors, i.e. bacterial chromosome mutations or mutations of pre-existing genes under different environmental pressures, can be positively correlated to drive the evolution of acquired mechanisms of antibiotic resistance in pathogenic and opportunistic bacteria. The reduced affinity of antibiotics for their targets may result from chromosomal mutations. Additionally, several mechanisms of antibiotic resistance such as chromosomal AmpC β-lactamases & efflux pumps are exceptionally controlled during their production and at a baseline level, offering a naturally reduced level of sensitivity to the antibiotics.

The mutations in the genomic structures that regulate sensitivity to antibiotics result in the overexpression of high levels of antibiotic resistance in them. However, lateral transfer from donor species speeds up the acquisition of antibiotic resistance genes due to adaptation to the selection pressure of antibiotics. All the naturally occurring and anthropogenically altered ecosystems share an underlying resistome, or baseline ARG level, thanks to the ubiquitous dispersion of ARGs. However, since the 1940s, during the onset of the era of antibiotics, human-controlled antibiotic inputs have made significant changes in the natural habitats and the environments where they are growing. The sul1 ARGs are associated with wastes from animal operations and upstream wastewater treatment plants. The anthropogenic activities have resulted in the appearance of "tet ARGs", "24 class 1 integrons", and "sul1 gene" in the Cuban Almendares River, River Thames in the United Kingdom, and Swiss lakes, respectively. Furthermore, Lupo et al. (2012) reported a 15-fold rise in the levels of 18 ARGs encoding for resistance to large spectrum beta-lactamases, erythromycins, and tetracyclines in the archival Dutch soils spanning from the pre- and post-antibiotic period.

9.4.4 Specific Examples of Wastewater Being the Source of Antibiotic Resistance

Antibiotic resistance is induced by a transitory shift in the equilibrium of resistant and susceptible strains in a bacterial population, which may be seen by phenotypic susceptibility. This is how certain microbes develop resistance to one or two antibiotics, as well as many others. When germs acquire resistance towards antibiotics, the drug treatment becomes futile and causes an elevation of illnesses to dangerous levels. Thus, it causes an increase in hospital treatment expenses, longer duration of treatment unless an effective drug is found, and subsequently the risk of spreading to other individuals (Hutchings et al., 2019) Antibiotic-resistant bacteria pose a serious threat to humans globally, emerging as a global threat and challenge to world. WHO published a "priority infections" list that includes MDR, highly drug-resistant (HDR), or resistant to the action of methicillin. Since these pathogenic organisms showing resistance to wide range of antibiotics pose a significant danger to human and animal health, new medications/drugs are required against these bacteria to control infections (Collignon et al., 2016).

Several microbes have evolved resistance to medications that have been accessible since the 2000s. The infections *Enterococcus faecium*, *Staphylococcus aureus*, *Klebsiella pneumoniae*, *Acinetobacter baumannii*, *Pseudomonas aeruginosa*, and *Enterobacter* species microorganisms (ESKAPE) are challenging to treat due to their high AMR. These microbes travel across environmental compartments and are the topic of a variety of studies (Bair et al., 2020).

9.4.4.1 *Salmonella* sp.

Research analyzed the features of food and human isolates to classify the source of infection by determining the incidence of the presence of *Salmonella* in sewage influent of wastewater treatment plants (WWTPs). For three years, two WWTPs in Yamanashi Prefecture, Japan, were the sites for collection of sewage influent samples every week. These findings imply that sewage influent contains human *Salmonella* isolates. The work also reported that these *Salmonella* isolates originated from several poultry and unreported human cases (Yanagimoto et al., 2020).

Typhoid fever is brought on by the Gram-negative, motile, rod-shaped, facultative anaerobe *Salmonella sp.* The following mechanisms describe how *S. typhi* avoids the effects of antibiotics: antibacterial agent inactivation, antimicrobial efflux or transport, the antibacterial target site being altered, and a decrease in the antibacterial agent's permeability. Most medication resistance is genetic in nature, either from chromosomal mutations, plasmid or transposon acquisition, or both. In some strains of *S. typhi*, plasmids of incompatibility group HI1 and C are significant carriers of antibiotic Resistance. Recent reports have linked a single point mutation in the DNA gyrase-encoding topoisomerase gene gyrA's quinolone resistance determination region (QRDR) to the chromosomally-mediated drug resistance phenomenon against fluoroquinolones (Ugboko & De, 2014).

9.4.4.2 Staphylococcus aureus

Staphylococcus spp. strains isolated from activated sludge (AS) bioreactor, river water collected upstream and downstream from the wastewater discharge point (URW and DRW), untreated (UWW), treated (TWW), and WWTP workers were the subjects of a study. This study aimed at discovering the genetic drivers of virulence and drug resistance in these organisms. Out of the total strains isolated from water samples and wastewater, around 63% of the isolated strains were identified as *S. aureus*, and the vanA gene was identified in 20% of the strains. The hla and pvl virulence genes were found in 80% and 27% of the isolates, respectively. The presence of mecA and van A genes was confirmed in 16% and 14% of the isolates, respectively, in the group of strains recovered from the employees. Hla was the most common virulence gene (74%), whereas pvl was found in 43% of the isolates (Zieliński et al., 2020).

The adaptive evolution among bacteria was best demonstrated by *Staphylococcus aureus* in the era of antibiotics better than any other known human pathogen, as it has proven to have the unusual potential to generate a resistance mechanism to every new antibiotic shortly, beginning with methicillin and penicillin, progressing to the most recent, daptomycin and linezolid (Pantosti and Sanchini, 2007).

Penicillinase and aminoglycoside-modification enzymes hydrolyze the amide bond of the β-lactam ring of ampicillin and penicillin to produce an inactive compound. The alteration in the target which causes decreased affinity for the antibiotic (e.g. penicillin-binding protein 2a) is the other known resistance mechanism (Pantosti and Sanchini, 2007).

The trapped vancomycin clogs the cell membrane, making vancomycin penetration even more difficult, antibiotic trapping (for vancomycin and probably daptomycin), and efflux pumps (fluoroquinolones and tetracycline). Resistance results from spontaneous mutations that alter the amino acids of either one or both of the topoisomerase IV and DNA gyrase enzymes, which are crucial for the replication of DNA. Point mutations in the component ParC of topoisomerase IV and subunit GyrB of DNA gyrase result in imparting *S. aureus* its resistance (Pantosti and Sanchini, 2007).

9.4.4.3 *Acinetobacter* spp.

Acinetobacter spp. strains were found in two receiving bodies and three wastewater treatment facility outputs, i.e. raw influent, second and final effluent, upstream and downstream of the treated point of wastewater discharge. Resistance to Rifampin, Amoxicillin, chloramphenicol, and multi-drug antibiotics in *Acinetobacter* isolates (Zhang et al., 2009).

Acinetobacter species are Gram-negative, non-fastidious, non-motile, aerobic, glucose-non-fermentative, oxidative-negative, catalase-positive and coccobacilli. Numerous studies have demonstrated that *A. baumannii* quickly acquires antimicrobial drug resistance, and multi-drug-resistant strains have been discovered (Boucher et al., 2009). Due to its remarkable capacity to develop resistance against antibiotics, *Acinetobacter* sp. has emerged as one of the pathogens with the greatest success in contemporary healthcare.

A. baumannii strains vary in their level of sensitivity to the majority of therapeutically available antibiotics (Lin & Lan, 2014). It possesses multiple resistance mechanisms, including β-lactamases, enzymes that transform aminoglycosides, permeability flaws, efflux pumps, and target site modifications (Gordon & Wareham, 2010; Lin & Lan, 2014). The number of antibiotic classes available to treat *Acinetobacter spp.* infections in clinical practice has continuously reduced due to the development of several resistance mechanisms in *Acinetobacter* spp. (Lee et al., 2017).

9.4.4.4 *Campylobacter* sp.

Enrichment stage and PCR-based studies were performed with the objective to assess the prevalence of pathogenic *Escherichia coli* O157:H7, *Campylobacter*, *E. coli* virulence genes, and *Salmonella* spp. in various WWTPs. PCR amplification suggested 8% and 25% of the effluents consist of *Campylobacter jejuni* and *E. coli* DNA, respectively. Whereas, influents have a higher percentage of pathogens consisting of 50% of *Campylobacter jejuni* and 25% of *E. coli* DNA.

Campylobacter is small (0.2–0.9μm broad and 0.2–5.0μm long), spirally curled, motile Gram-negative bacteria found in the intestines of wild and domestic animals. In developed countries, campylobacteriosis is the most common gastroenteric infection in humans that is caused by *Campylobacter* bacteria (Blaser and Engberg, 2008).

Globally, each year an estimated 400–500 million people suffer from diarrhoea caused by *Campylobacter*. Humans are typically infected by zoonotic *Campylobacter* species, primarily *C. jejuni*, and *E. coli* that enter the human body by consumption of contaminated water or food (Humphrey et al., 2007; Altekruse and Tollefson, 2003).

Quinolones cause bacterial cell death by preventing the production of bacterial DNA. DNA gyrase and topoisomerase IV are two significant bacterial enzymes that quinolones target. These enzymes collaborate to replicate, transcribe, recombine, and repair DNA in bacteria. Other mechanisms of this resistance exist as well, such as an efflux system and reduced outer membrane permeability. GyrB and GyrA (i.e. DNA gyrase) and ParE and ParC (i.e. topoisomerase IV) are two pairs of subunits that make up the large enzymatic quaternary structures known as gyrase and topoisomerase. The major cause of resistance to fluoroquinolones is an amino acid substitution(s) in the topoisomerase's quinolone resistance-determining region (QRDR) (Wieczorek et al., 2013).

9.4.4.5 *Enterococcus* sp.

In two WWTPs in the same municipality – a conventional activated sludge (CAS) and a biologically aerated filter (BAF) – vancomycin-resistant *Enterococci* (VRE) were utilized as indicators of antimicrobial resistance (AMR). The primary and final effluents were counted for *Enterococci* and VRE (Sanderson et al., 2019).

Gram-positive, facultative anaerobic *Enterococci* are known for their capacity to hydrolyze bile-esculin and L-pyrrolidonyl-B-naphthylamide. They grow at high pH and (6.5%) NaCl concentrations (PYR). All therapeutically relevant antibiotics are potentially resistant to *Enterococci*. Their rise to prominence as significant nosocomial infections and the genus's growing expression of antibiotic resistance are related. The mechanisms underpinning *Enterococci's* antibiotic resistance might be inherent to the species. These are acquired by mutation of intrinsic genes or transfer of genetic material encoding resistance through HGT. The majority of bacterial cell walls include peptidoglycans, which are linked together by enzymes to form pentapeptide precursor molecules.

PBPs are the enzymes involved in these cross-linking events because β-lactams (structural analogues of pentapeptide precursors) bind covalently and prevent proper cell wall development. Reactive oxygen species are produced when β-lactam agents attach to PBPs, which usually leads to poor cell wall production and programmed cell death. Some *Enterococci* have low-affinity PBPs that weakly bind β-lactam antibiotics (PBP4 and PBP5 in *E. faecium*, and *E. faecalis*, respectively). Due to activity brought on by lsa gene expression, *E. faecalis* is inherently resistant to clindamycin (a lincosamide), dalfopristin (streptogramin A class), and quinupristin (streptogramin B class). Drug efflux is suggested as a potential mechanism due to structural similarities between lsa and ABC efflux pumps. The antibiotic combination trimethoprim-sulfamethoxazole inhibits folate synthesis and kills a variety of bacterial species simultaneously by blocking two consecutive stages in the tetrahydrofolate synthesis pathway. Since they can take folic acid from their surroundings and avoid the effects of trimethoprim-sulfamethoxazole, *Enterococci* are exceptional in this regard (Hollenbeck & Rice, 2012).

9.4.4.6 *Klebsiella* sp.

Wastewater acts as a significant means for the spread of antibiotic resistance because after ingestion, exposed bacteria and drug residues end up there through faeces. Antimicrobial intake in the various sites, which was highest in the hospital, was matched in the bacterial AMR profiles. However, for all antimicrobials evaluated, less than 10% of the AMR reported in the WWTP was caused by effluents from the hospital (Verburg et al., 2019).

The World Health Organization, the US Centres for Disease Control and Prevention, and the UK Department of Health have all designated that *Klebsiella pneumoniae* is an immediate threat to human health due to its MDR nature. Illnesses with *K. pneumoniae* are particularly problematic for newborns, the elderly, and those with impaired immune systems in the healthcare environment. However, this bacterium is also responsible for a sizeable proportion of community-acquired infections including pneumonia and sepsis. Colistin resistance in MDR *K. pneumoniae* has been linked to loss-of-function mutations in the mgrB gene, according to several studies. In *K. pneumoniae*, the mgrB mutation is linked to lipid A remodelling under PhoPQ control, which confers resistance to polymyxins and mammalian antimicrobial peptides. The resistance to polymyxins is mediated by lipid A alterations with 2-hydroxymyristate and Ara4N (Kidd et al., 2017). By producing enzymes like Extended Spectrum β-Lactamases (ESBLs) and Carbapenemase, *K. pneumoniae* is more likely than other bacteria to become resistant to antibiotics (Nirwati et al., 2019). By increasing the affinity between *Klebsiella pneumoniae* and penicillin, PBPs prevent *Klebsiella pneumoniae* from growing. Importantly, zinc finger nuclease targets the downregulation of the β-lactam resistance gene that controls the manufacturing of β-lactamase. Two major causes of β-lactam antibiotic Resistance are (i) target mutation of gene-coding PBPs, and (ii) ESBLs. The indiscriminate use of antibiotics increases ESBL expression, which contributes to β-lactam antibiotic resistance via hydrolysis. β-lactam resistance in *Klebsiella pneumoniae* is accelerated by the hydrolysis of β-lactam antibiotics to produce antimicrobial inactive chemicals, β-lactamases (Jiang et al., 2020).

9.4.4.7 Pseudomonas aeruginosa

In wastewater influents, *Pseudomonas* was found to be the most prevalent species associated with ESKAPE, accounting for 21.356% of the sequence reads. *Streptococcus* was found to be the next common bacterium with 6.445% of the sequence reads. The other bacteria are *Acinetobacter Enterococcus, Klebsiella, Escherichia*, and *Staphylococcus* with 0.968%, 0.063%, 0.038%, 0.028%, and 0.028% and 0.004%, respectively. The *Escherichia*, *Klebsiella*, and *Acinetobacter* from the effluent were resistant to 11 different antibiotics. *Streptococcus* and *Staphylococcus* were resistant to nine different antibiotics, while *Pseudomonas* was resistant to seven different antibiotics (Limayem et al., 2019).

More than 100 distinct species of the non-fermenting Gram-negative *Pseudomonas* genus have been identified since Emil (Walter) Migula, a Polish-born German botanist, originally named and categorized it in 1894 (Palleroni, 2003). *Pseudomonas* genus is considered to be one of the most diverse and widespread groups. Members of the *Pseudomonas* group are isolated from a variety of artificial, natural (e.g. soil, waters), and clinical habitats. They have rapid growth and adaptability to a variety of environmental conditions (oxidative, other stresses, and nutritional) (e.g. contact lenses, catheters, activated sludge).

The ability to produce a diverse range of extracellular products is a key characteristic of *Pseudomonas* species. Extracellular polymeric molecules, such as proteins, polysaccharides, enzymes, and extracellular DNA, have been implicated in several activities, primarily attachment procedures, biofilm development, and pathogenicity. Furthermore, *Pseudomonas* species release biosurfactants that promote the solubilization of hydrocarbons from non-aqueous phase liquids and make them accessible for microbial breakdown (van Delden, 2004).

As a result, *Pseudomonas spp.* are capable of mineralizing a variety of organic substances, including chloro- and nitro-organic compounds, aromatic hydrocarbons, herbicides, and pesticides. This is essential for the detoxification and bioremediation of polluted environments. *Pseudomonas* species are crucial contributors to the microbial population involved in the methods used to remediate wastewater. They contain gene clusters related to denitrification as well as those needed for polyphosphate (polyP) production (polyphosphate kinase ppk) and degradation (exopolyphosphatase ppx) (Tobin et al., 2007).

P. aeruginosa is linked to gastrointestinal, joint, bone, urinary, soft tissue, and surgical site infections in addition to respiratory tract infections. Chronic and acute *Pseudomonas* infections are frequently difficult to treat. Along with the capacity to build biofilms and thrive in anoxic environments, *Pseudomonas* spp. also exhibit low sensitivity to a variety of antimicrobials and disinfectants. For *Pseudomonas* spp., co-occurring mechanisms such as reduced outer membrane permeability, the production of β-lactamases, and efflux systems are the principal sources of natural (intrinsic) resistance. Because of the amazing flexibility of the genome, the members of the *Pseudomonas* genus show all the known antimicrobial-resistant mechanisms. Since there are few effective treatments available, *Pseudomonas* infections are frequently linked to multi-drug resistance (Luczkiewicz et al., 2015).

9.4.4.8 *Enterobacter* sp.

According to a study, the *Enterobacteriaceae* found in hospitals, slaughterhouses, and downstream water bodies are resistant to conventional antimicrobials. More MDR bacteria were present in hospital effluents, which posed a serious risk to the public's health by migrating to the water bodies farther downstream (Tesfaye et al., 2019).

The primary mechanism of beta-lactam resistance in *Enterobacteriaceae* is β-lactamase synthesis. The periplasmic space is where these extremely diverse enzymes hydrolyze beta-lactams, avoiding penicillin-binding protein inhibition. Typically, the intrinsic beta-lactamase content of *Enterobacteriaceae* is used to classify them. In particular, *Citrobacter freundii*, *Enterobacter* sp., *Hafnia alvei*, *Morganella morganii*, *Providencia* sp., and *Serratia marcescens* generate a chromosome-encoded, inducible, AmpC cephalosporinase. Amoxicillin, first-generation cephalosporins (1GC), cefoxitin, and clavulanic acid all generate AmpC, which causes intrinsic resistance. While other beta-lactams are lesser inducers, carbapenems are similarly effective inducers but continue to be active because of the lack of considerable AmpC-mediated hydrolysis.

The major source of aminoglycoside resistance in *Enterobacteriaceae* is AMEs, which reduce antibiotic action by engrafting different radicals (APH, aminoglycoside nucleotidyltransferase, AAC, aminoglycoside phosphotransferase, ANT, and aminoglycoside acetyltransferase). More recently, aminoglycosides' targets, the 16S ribosomal subunit methylases, have been identified, particularly in bacteria that produce NDM. These enzymes, Rmt, and ArmA save neomycin by conferring resistance to all aminoglycosides. Quinolones and fluoroquinolones are inherently toxic to all *Enterobacteriaceae*. After several chromosomal mutations in the topoisomerase IV and DNA gyrase-encoding genes (parC and gyrA, respectively), leading to an increase in the MICs with each mutation, high-level resistance develops. As a result, organisms with a single mutation may look extremely resistant to quinolones while being vulnerable to fluoroquinolones. Chromosomal mutations may potentially result in lower permeability or efflux pump overexpression, leading to diminished susceptibility. In addition to mutations, plasmid-encoded resistance with Qnr (A, B, C, D, and S subtypes) emerged in the 2000s. Also, a low-level fluoroquinolone resistance, which is conferred by small DNA-mimicking protein, AAC(6′)-Ib-cr, an AME that extends the ciprofloxacin, norfloxacin, and QepA resistance spectrum because of two mutations (Ruppé et al., 2015) (Table 9.2).

9.5 METHODS FOR MITIGATION OF ANTIBIOTIC RESISTANCE GENERATED THROUGH WASTEWATER

A growing, all-encompassing concern to world health is AMR, which arises as a threat to advancements in treatment of contagious illnesses achieved in the last decades. Antimicrobial resistance is a complicated issue with several linked variables, some of which may be shifting patterns of commerce, population growth, and climate change. To overcome this issue and to assist international and national priority settings, treatment decisions, and public health initiatives, reliable information that precisely specifies and characterizes the worldwide incidence and transmission of AMR is crucial.

Understanding the effectiveness and workings of the treatment methods, as well as the constraints on their future use, is crucial for reducing the effects of ARB and ARGs on the environment and human health. Adsorption, biodegradation, disinfection, and filtering employing membranes with narrow pore sizes are some removal processes. Depending on the characteristics of the antibiotic, other routes such as photolysis, volatilization, and hydrolysis, contribute to elimination (Bair et al., 2020).

9.5.1 Anaerobic, Aerobic, or Combined Treatments

Chemical oxygen demand (COD) is often treated using aerobic and anaerobic treatment methods since they use less energy and are ecologically beneficial. They may also effectively eliminate ARB and ARGs. Aerobic treatments occur in the presence of air where organic pollutants are converted into carbon dioxide, water, and biomass in the presence of aerobic microorganisms that can utilize oxygen. On the other hand, during anaerobic procedures, pollutants are converted into carbon

TABLE 9.2
Examples of Microorganisms Found in Wastewater Habitats

Resistant Microbial Strain	Profile of Resistance	Sample Site	References
Salmonella sp.	Kanamycin, streptomycin, tetracycline	Sewage influent of WWTPs, Japan	(Yanagimoto et al., 2020)
Staphylococcus aureus	Vancomycin, methicillin	Activated sludge bioreactor Olsztyn, Poland	(Zielin´ski et al., 2020)
Acinetobacter spp.	Rifampin, amoxicillin, chloramphenicol	Effluent and Influent from WWTPs and receiving body of the plant (River), Michigan USA	(Zhang et al., 2009)
Campylobacter	Ciprofloxacin, gentamicin, erythromycin, nalidixic acid, streptomycin	Isolated from Poultry Carcasses in Poland	(Wieczorek et al. , 2013)
Enterococcus	Multi-resistant	Primary and final effluent, and biomass, Canada	(Sanderson et al., 2019)
Klebsiella spp.	Ampicillin, cotrimoxazole, ciprofloxacin, and trimethoprim	Affluent and effluent from the WWTP plant, City of Sneek, The Netherlands	(Verburg, et al., 2019)
Pseudomonas	Multi-resistant	Affluent and effluent, Florida	(Limayem et al., 2019)
Enterobacter sp.	Ciprofloxacin, penicillin and norfloxacin	Effluent, China	(Tesfaye et al., 2019)

dioxide and methane in the presence of anaerobic microorganisms that do not mandatorily require oxygen (Barancheshme & Munir, 2018).

The main methods for removing antibiotics, antibiotic resistance (ABR) organisms, and ABR genes from different biological processes are thought to be biosorption and biodegradation. When aerobic and anaerobic processes are combined throughout the wastewater treatment process, performance is improved. Other research has demonstrated that thermophilic anaerobic digestion promotes the elimination of numerous ARGs for macrolide resistance genes by up to 89% at temperatures between 40 and 60°C (Christgen et al., 2015). Not all ARGs, nevertheless, are likely to be decreased by anaerobic therapies. During anaerobic digestion, ARGs from tetracycline and erythromycin increased. Studies demonstrated that anaerobic digestion can be combined with microwave, heat hydrolysis, and ozone for improved outcomes in certain situations (Aydin et al., 2016; Bairán et al., 2020).

The cell membrane is harmed by microwaves, which disrupt DNA. DNA decrease results from the sterilization of sludge microorganisms and the destruction of their cell walls during thermal hydrolysis (Xue et al., 2019). If ozone can enter the cytoplasm to cause gene decrease, the nonselective oxidation of ozone may also be effective. Less research has been done on the elimination of ARGs by aerobic digestion. In aerobic, anaerobic, and mixed settings, antibiotics such as macrolides, sulfonamides, and tetracyclines have been eliminated. For instance, three sewage treatment facilities in Xinjiang had removal rates for five sulfonamides, five quinolones, and four tetracyclines that ranged from 17% to 100%.

Membrane bioreactors (MBRs) are bioprocess units that employ a membrane with a regulated and fine pore size in amalgamation with the principles of mass separation based on chemical affinity and/or molecular size. In a sophisticated filtration process, the membrane functions as a barrier to remove undesirable substances from mixtures of liquid or gas. ARB and ARGs may now be effectively removed using membrane bioreactors. For instance, an anaerobic fluidized bed reactor is linked externally to a ceramic membrane (0.05 μm pore size), which results in production of a 3.3–3.6 log decrease of tet(O), erm(B), and sul1, along with the HGT determinant intI1 (Kappell et al., 2018).

ARGs (ermB, intI1, tetO, and tetW), antibiotics (ampicillin, ciprofloxacin, tetracycline, erythromycin, and kanamycin), ARBs (*Aeromonas*, *Bacteroides*, *Klebsiella*, and *Escherichia*), and ARBs (*Aeromonas*, *Escherichia*, and *Klebsiella*). The findings demonstrated the high rates of approximately 94% and 77.5% for removal of ARBs and degradation of all antibiotics, respectively. The study also demonstrated that MBRs can help in removal three to four orders of magnitude higher ARGs that could be removed by other methods (Bair et al., 2020).

9.5.2 Constructed Wetlands

The tiny semi-aquatic habitats that provide a suitable environment for the existence of a variety of microbial communities, with simultaneous presence of several physico-chemical interactions, are called constructed wetlands (Barancheshme & Munir, 2018).

Constructed wetlands (CWs) are created to simulate natural water-purification processes. Domestic, livestock, and municipal wastewater have all been treated using them. It has been shown that these systems are capable of reducing emergent pollutants of various origins, as well as the chemical oxygen demand (COD), biochemical oxygen demand (BOD), phosphorus, and nitrogen. It is feasible to build specific configurations for removing certain contaminants from effluents by operating CWs under various macrophytes, flow configurations, and solid supports (Bair et al., 2020) Biodegradation is the most important mechanism in the elimination of these pollutants. Plant uptake and substrate adsorption are also key to reducing the nutrient loadings, ARGs, and antibiotics in the artificial wetlands (Barancheshme & Munir, 2018). Constructed wetlands can effectively eliminate aqueous ARGs, yet they can also serve as a reservoir for certain ARGs. Eight antibiotics, 12 genes, and 16S rRNA (bacteria) were screened in various combinations to investigate the rate of ARB and ARG removal from the domestic wastewater setups in specially designed different types of wetlands. The designed wetlands under different combinations showed different results with aqueous removal efficiency of total antibiotics and total ARGs from 75.8% to 98.6% and 63.9% to 84.0%, respectively (Barancheshme & Munir, 2018) In CWs, there is a variety of ARB and ARG removal percentages. In the case of antibiotics such as clarithromycin, erythromycin, leucomycin, monensin, sulfamethoxazole, sulfamethazine, sulfapyridine, and trimethoprim (Bair et al., 2020).

9.5.3 Disinfection

A substantial number of harmful organisms such as bacteria, viruses, or parasites that cause infections are killed during the disinfection of water/wastewater. Chlorination is one of the most widely used disinfection techniques for wastewater treatment because of its ease of application and efficiency. However, UV radiation and ozone exposure are also used (Barancheshme & Munir, 2018). Effective disinfection is a crucial and ongoing procedure for the inactivation of dangerous pathogens as well as the disruption of ARB and ARGs. To minimize harmful bacteria, water disinfection is often used after the conclusion of the wastewater treatment process. Inactivation of pathogens is the primary goal of disinfection. But more recently, one key concern of disinfection is the elimination of antibiotic Resistance genes that confer resistance to bacteria against antibiotics. Chlorination, ozonation, and UV radiation are three important disinfection techniques that can be used alone or in combination (Bair et al., 2020). For ozonation, the decrease of ARGs (ampC, blaVIM, ermB, and vanA) varied from 18.7–99.3%. The eradication of ARBs (*Enterobacteria*, *Enterococci*, *P. aeruginosa*, and *Staphylococci*) ranged from 60.2–98.9% (Bairán et al., 2020). In another study, following ozone treatment, ARGs, i.e. blaVIM, and vanA, increased within the existing wastewater population whereas a decrease by two orders of magnitude was observed for erythromycin resistance gene (ermB) (Alexander et al., 2016).

Chlorination's impact on the elimination of several antibiotics, including erythromycin, ciprofloxacin, rifampicin, tetracycline, chloramphenicol, gentamicin, sulfadiazine, cephalexin, and cephalexin and vancomycin as well as its impact on the inactivation of ARB and ARGs, has previously been studied. Advanced oxidation techniques have also been used in disinfection procedures, including UV radiation, ozonation, photocatalytic systems, and Fenton reagents (Barancheshme & Munir, 2018).

A comparative study evaluated the efficiency of three disinfection techniques: single UV radiation, chlorination, and sequential UV/chlorination, in order to combat ARGs in municipal WWTPs (Sharma et al., 2016). The study demonstrated that a positive correlation exists between the maximal inactivation of an ARG, specifically intI(1), 16S rRNA, sul(1), tet(G), and tet(X), and the chlorine dose and contact duration. The results showed that at 30 mg min. L of Cl_2, the maximal inactivation ranged from 1.30-log to 1.49-log. The best removal efficiency was exhibited by sequential UV/chlorination for ARGs, while single UV exposure was less efficient.

9.5.4 Nanomaterial

Antimicrobial nanotechnology has demonstrated that it is an efficient barrier against ABR and ARGs through the use of several unique combinations of nanomaterials. When nanoparticles cure antibiotic resistance, there are likely two different processes at work. In the first method, the coupling of a functionalized nanomaterial with antibiotics is performed. When this antibiotic-rich functionalized nanomaterial enters the AR, it results in the release of a significant number of harmful ions against ARGs. The second technique employs a mix of antibiotics and nanomaterials to attack ARGs in a complementary manner (Barancheshme & Munir, 2018). Due to their antibacterial action, nanomaterials can be regarded as a barrier against MDR. The two key factors that determine the antibacterial activity of nanoparticles are the physicochemical characteristics of nanoparticles and the kind of target bacteria (Hajipour et al., 2012). Although there is a significant correlation in a few facets of the antimicrobial effect of nanoparticles, findings are difficult to generalize because most researchers conduct experiments with bacteria and nanoparticles that are readily available rather than specifically targeting desired bacteria or nanoparticles. It is challenging to relate the nanoparticles employed in lab investigations to fundamental physicochemical characteristics for full-scale manufacture (Barancheshme & Munir, 2018). Nanobiocides were employed by researchers to identify antimicrobial nanoparticles and categorize them as carbon-based, metal oxide, and natural nanomaterials. Owing to their strong antibacterial activity, the nanomaterials are employed in air filters, water treatment facilities, and medical procedures. Silver nanoparticles are the most studied/employed materials for water treatment procedures, either alone or together with biodegradable polymeric networks. Gram-positive, Gram-negative, and viral inactivation have been demonstrated for other metals such as copper, TiO2, and zinc. A few of these compounds, in addition to having strong antibacterial properties, are also affordable, water-resistant, and may be integrated into thin films or nanotubes, depending on the application (Bair et al., 2020) Activated carbon nanoparticles made from biowaste showed antifungal or antibacterial activity similar to functionalized polymeric materials, which may be helpful in water-purification procedures. Despite all of their antibacterial qualities, nanoparticles can promote antibiotic resistance, could be harmful to mammalian cells, and can cause ARGs. Nanoparticles have been utilized in the evaluation of harmful bacterial resistance to antibiotics. The formation of biofilms or the spread of genes that make bacteria resistant to metals are two additional ways that bacteria might acquire metal resistance. In other words, even if bacteria or pathogens are eliminated, their genes may still spread to new bacteria or microbes, increasing the likelihood that they may develop a resistance to antibiotics (Bair et al., 2020).

9.5.5 Coagulation

Coagulation is an active process that treats turbidity, colour, natural organic matter, and heavy metals in addition to removing colloidal particles from water. While coagulants have a positive electrical charge, colloidal particles often have a negative electrical charge. Particles are neutralized by a coagulant, which causes them to adhere when they come into contact. In WWTPs, the coagulation process is frequently used as the tertiary treatment step to improve water quality and remove impurities. According to the target pollutants, many kinds of organic and inorganic coagulants have been developed. Research has been done to examine how well zirconium coagulant removes natural organic debris from drinking water. Zirconium coagulant has enhanced water quality in comparison to iron coagulant, and also offers decreased dissolved organic carbon residue (Jarvis et al., 2012).

The removal effectiveness can be impacted by the coagulant dose, solution pH, coagulants, turbidity, natural organic matter content, and flocculation time. Recently, researchers have looked at how well coagulation technology removes ARGs from treated wastewater. The active process for removing ARGs from WWTP effluent is coagulation. Researchers investigated the elimination of integrase genes, tet genes, and sul genes using inorganic coagulant $FeCl_3$ and inorganic polymer coagulant poly ferric chloride (PFC). Significant relationships between dissolved NH_3-N, DOC1, and ARGs were found. The effectiveness of ARG removal ranged from 0.5-log to 3.1-log reductions (Barancheshme & Munir, 2018) (Table 9.3).

9.6 CONCLUSION AND PROSPECTS

The advent of bacterial species on our planet occurred billions of years ago, whereas the application of antimicrobials in therapeutic treatment is a recent historical development. The rapid development of antibiotic resistance among bacteria is a threat to the development of next-generation treatment techniques. The need for cutting-edge treatment methods grows as antibiotic resistance among diseases emerges. There are still many gaps to be filled despite the substantial efforts that have been made to research the therapeutic options for ARB and ARGs. Untreatable infections have increased in frequency in clinical settings due to the infected organism's multi-drug resistance. The lack of research and development on antibiotics has exacerbated this critical situation. Research and development activities need to be significantly strengthened and encouraged if we are to successfully address this issue. It is crucial to have a thorough understanding of the mechanisms by which

TABLE 9.3
Treatment Processes for Antimicrobial Resistance Mitigation

Resistant Microbial Strain	Treatment	Country	References
Escherichia coli	Disinfection and solar-driven advanced oxidation processes	Italy	(Fiorentino et al., 2015)
Bacteroides, Enterococcus, Ferruginibacter, Morganella, Mycobacterium, Myroides, Romboutsia, Thermomonas	Combined UV and chlorine process	China	(Wang et al., 2020)
Escherichia coli NDM-1	Nitrogen-doped reduced graphene oxide (NRGO), photocatalytic process	China	(Li et al., 2020)
Escherichia coli and P. aeruginosa	Electrochemical	China	(Chen et al., 2020)
Staphylococcus aureus, Escherichia coli, Klebsiella pneumoniae	Solar light and solar photo-Fenton processes	Switzerland	(Giannakis et al., 2018)

bacteria develop antibiotic resistance in order to develop tactics to deal with the threat of resistance. Antibiotics must be created with the knowledge that bacteria will react to them and develop resistance to them (an evolutionary fact). Therefore, ongoing, tenacious, and constant efforts should be made to create antibiotics and investigate the causes of resistance.

CONFLICT OF INTEREST

All the authors approve of the submission and do not have any conflicts of interest to declare.

ACKNOWLEDGEMENT

BK wishes to acknowledge the DST-SERB National Post-Doctoral Fellowship (PDF/2022/001781) for financial support.

REFERENCES

Alexander, J., Knopp, G., Dötsch, A., Wieland, A., & Schwartz, T. (2016). Ozone treatment of conditioned wastewater selects antibiotic resistance genes, opportunistic bacteria, and induce strong population shifts. *Science of The Total Environment*, 559:103–112. https://doi.org/10.1016/j.scitotenv.2016.03.154

Altekruse, S.F., & Tollefson, L.K. (2003). Human campylobacteriosis: a challenge for the veterinary profession . *Journal of the American Veterinary Medical Association,* 223(4):445–452. https://doi.org/10.2460/javma.2003.223.445

Aydin, S., Ince, B., & Ince, O. (2016). Assessment of anaerobic bacterial diversity and its effects on anaerobic system stability and the occurrence of antibiotic resistance genes. *Bioresources Technology*, 207, 332–338.

Bairán, G., Rebollar-Pérez, G., Chávez-Bravo, E., & Torres, E. (2020). Treatment processes for microbial resistance mitigation: The technological contribution to tackle the problem of antibiotic resistance. *International Journal of Environmental Research and Public Health*, 17(23). https://doi.org/10.3390/ijerph17238866

Bair, G., Rebollar-P, G., Ch, E., & Torres, E. (2020). Treatment processes for microbial resistance mitigation: The technological contribution to tackle the problem of antibiotic resistance. *International Journal of Environmental Research of Public Health*, 17:8866. https://doi.org/10.3390/ijerph17238866

Barancheshme, F., & Munir, M. (2018, January 8). Strategies to combat antibiotic Resistance in the wastewater treatment plants. *Frontiers of Microbiology*, 8:2603. https://doi.org/10.3389/fmicb.2017.02603

Blair, J.M., Richmond, G.E., & Piddock, L.J. (2014). Multidrug efflux pumps in Gram-negative bacteria and their role in antibiotic resistance. *Future Microbiology*, 9:1165–1177.

Blaser, B., & Engberg, J. (2008). Clinical aspects of *Campylobacter jejuni* and *Campylobacter coli* infections. In Nachamkin, C., Szymanski, M., & Blaser, M.J. (eds.) *Campylobacter.* American Society for Microbiology, Washington, DC, pp. 99–121.

Boczek, L.A., Rice, E.W., Johnston, B., & Johnson, J.R. (2007). Occurrence of antibiotic-resistant uropathogenic *Escherichia coli* clonal group a in wastewater effluents. *Applied and Environmental Microbiology,* 73(13):4180–4184. https://doi.org/10.1128/AEM.02225-06

Boucher, H.W., Talbot, G.H., Bradley, J.S., Edwards, J.E., Gilbert, D., Rice, L.B., et al. (2009). Bad bugs, no drugs: no ESKAPE! An update from the Infectious Diseases Society of America. *Clinical Infectious Disease,* 48(1):1– 12. https://doi.org/10.1086/595011

Cetinkaya, Y., Falk, P., & Mayhall, C.G. (2000). Vancomycin-resistant enterococci. *Clinical Microbiology Reviews*, 13(4):686–707. https://doi.org/10.1128/cmr.13.4.686-707.2000

Chen, L., Zhou, Z., Shen, C., & Xu, Y. (2020). Inactivation of antibiotic-resistant bacteria and antibiotic resistance genes by electrochemical oxidation/electro-Fenton process. *Water Science & Technology,* 81(10):2221–2231. https://doi.org/10.1016/j.cej.2023.141807

Christgen, B., Yang, Y., Ahammad, S.Z., Li, B., Rodriquez, D.C., Zhang, T., & Graham, D.W. (2015). Metagenomics shows that low-energy anaerobic-aerobic treatment reactors reduce antibiotic resistance gene levels from domestic wastewater. *Environmental Science & Technology*, 49:2577–2584. https://doi.org/10.1021/es505521w

Collignon, P.C., Conly, J.M., Andremont, A., McEwen, S.A., Aidara-Kane, A., Agerso, Y., Andremont, A., Conly, J., WHO-AGISAR, Agerso, Y., Andremont, A., Collignon, P., Conly, J., Ninh, T.D., Donado-Godoy, P., Fedorka-Cray, P., Fernandez, H., Galas, M., Irwin, R., Karp, B., Matar, G., McDermott, P., McEwen, S., Mitema, E., Reid-Smith, R., Scott, H.M., Singh, R., DeWaal, C.S., Stelling, J., Toleman, M., Watanabe, H., & Woo, G-J. (2016). World Health Organization ranking of antimicrobials according to their importance in human medicine: A critical step for developing risk management strategies to control antimicrobial resistance from food animal production. *Clinical Infectious Disease*, 63(8):1087–1093. https://doi.org/10.1093/cid/ciw475

da Silva, M.F., Tiago, I., Veríssimo, A., Boaventura, R.A.R., Nunes, O.C., & Manaia, C.M. (2006). Antibiotic resistance of *Enterococci* and related bacteria in an urban wastewater treatment plant. *FEMS Microbiology Ecology,* 55:322–329. https://doi.org/10.1111/j.1574-6941.2005.00032.x

Fiorentino, A., Ferro, G., Alferez, M.C., Polo-López, M.I., Fernández-Ibañez, P., & Rizzo, L. (2015). Inactivation and regrowth of multidrug resistant bacteria in urban wastewater after disinfection by solar-driven and chlorination processes. *Journal of Photochemistry and Photobiology B: Biology*, 148:43–50. https://doi.org/10.1016/j.jphotobiol.2015.03.029

Giannakis, S., Le, T.T.M., Entenza, J.M., & Pulgarin, C. (2018). Solar photo-Fenton disinfection of 11 antibiotic-resistant bacteria (ARB) and elimination of representative AR genes. Evidence that antibiotic resistance does not imply resistance to oxidative treatment. *Water Research*, 143:334–345. https://doi.org/10.1016/j.watres.2018.06.062

Girlich, D., Poirel, L., & Nordmann, P. (2010). Novel ambler class A Carbapenem-hydrolyzing β -lactamase from a *Pseudomonas fluorescens* isolate from the Seine river,, Paris , France. *Antimicrobial Agents and Chemotherapy*, 54(1):328–332. https://doi.org/10.1128/AAC.00961-09

Goni-Urriza, M., Capdepuy, M., Arpin, C., Raymond, N., Caumette, P., & Quentin, C. (2000). Impact of an urban effluent on antibiotic resistance of riverine *Enterobacteriaceae* and *Aeromonas* spp. *Applied and Environmental Microbiology*, 66:125–32. https://doi.org/10.1128/AEM.66.1.125-132.2000

Gordon, N.C., & Wareham, D.W. (2010). Multidrug-resistant *Acinetobacter baumannii*: Mechanisms of virulence and resistance. *International Journal of Antimicrobial Agents,* 35(3):219–226. https://doi.org/10.1016/j.ijantimicag.2009.10.024

Graf, F.E., Palm, M., Warringer, J., & Farewell, A. (2019). Inhibiting conjugation as a tool in the fight against antibiotic resistance. *Drug Development Research*, 80(1):19–23. https://doi.org/10.1002/ddr.21457

Grehs, B.W.N., Linton, M.A.O., Clasen, B., Oliveira, A.De., & Elvis, S. (2021). Antibiotic resistance in wastewater treatment plants: Understanding the problem and future perspectives. *Archives of Microbiology*, 203(3):1009–1020. https://doi.org/10.1007/s00203-020-02093-6

Grossman, T.H. (2016). Tetracycline antibiotics and resistance. *Cold Spring Harbor Perspectives in Medicine,* 6(4):025387. https://doi.org/10.1101/cshperspect.a025387

Haaber, J., Penades, J.R., & Ingmer, H. (2017). Transfer of antibiotic resistance in *Staphylococcus aureus. Trends in Microbiology*, 25:893–905. https://doi.org/10.1016/j.tim.2017. 05.011

Hajipour, M.J., Fromm, K.M., Ashkarran, A.A., Jimenez de Aberasturi, D., de Larramendi, I.R., Rojo, T., Serpooshan, V., Parak, W.J., & Mahmoudi, M. (2012). Antibacterial properties of nanoparticles. *Trends in Biotechnology*, 30(10):499–511. https://doi.org/10.1016/j.tibtech.2012.06.004

Hernando-Amado, S., Coque, T.M., Baquero, F., & Martinez, J.L. (2019). Defining and combating antibiotic resistance from one health and global health perspectives. *Nature Microbiology*, 4(9):1432–1442. https://doi.org/10.1038/s41564-019-0503-9

Hoffman-Roberts, H.L., Babcock, C.E., & Mitropoulos, I.F. (2005). Investigational new drugs for the treatment of resistant pneumococcal infections. *Expert Opinion on Investigational Drugs*, 14:973–995. https://doi.org/10.1517/13543784.14.8.973

Hollenbeck, B.L., & Rice, L.B. (2012). Intrinsic and acquired resistance mechanisms in *Enterococcus. Virulence*, 15(3, 5):421–433. https://doi.org/10.4161/viru.21282

Hong, P.Y., Al-Jassim, N., Ansari, M.I., & Mackie, R.I. (2013). Environmental and public health implications of water reuse: Antibiotics, antibiotic resistant bacteria, and antibiotic Resistance genes. *Antibiotics,* 2(3):367–399. https://doi.org/10.3390/antibiotics2030367

Hooper, D.C. (2002). Fluoroquinolone resistance among Gram-positive cocci. *The Lancet Infectious Disease*, 2(9):530–538. https://doi.org/10.1016/S1473-3099(02)00369-9

Humphrey, T., O'Brien, S., & Madsen, M. (2007). Campylobacters as zoonotic pathogens: A food production perspective. *International Journal of Food Microbiology*, 117(3):237–257. https://doi.org/10.1016/j.ijfoodmicro.2007.01.006

Hutchings, M.I., Truman, A.W., & Wilkinson, B. (2019). Antibiotics: Past, present, and future. *Current Opinion in Microbiology*, 51:72–80. https://doi.org/10.1016/j.mib.2019.10.008

Jalal, S., Ciofu, O., Høiby, N., Gotoh, N., & Wretlind, B. (2000). Molecular mechanisms of fluoroquinolone resistance in *Pseudomonas aeruginosa* isolates from cystic fibrosis patients. *Antimicrobial Agents and Chemotherapy*, 44(3), 710–712. https://doi.org/10.1128/aac.44.3.710-712.2000

Jarvis, P., Sharp, E., Pidou, M., Molinder, R., Parsons, S.A., & Jefferson, B. (2012). Comparison of coagulation performance and floc properties using a novel zirconium coagulant against traditional ferric and alum coagulants. *Water Research*, 46, 4179–4187. https://doi.org/10.1016/j.watres.2012.04.043

Jiang, W.E.I., Yang, W., Zhao, X., Wang, N.A., & Ren, H. (2020). *Klebsiella pneumoniae* presents antimicrobial drug resistance for β-lactam through the ESBL / PBP signaling pathway. *Experimental and Therapeutic Medicine* 19:2449–2456. https://doi.org/10.3892/etm.2020.8498

Kappell, A.D., Kimbell, L.K., Seib, M.D., Carey, D.E., Choi, M.J., Kalayil, T., Fujimoto, M., Zitomer, D.H., & McNamara, P.J. (2018). Removal of antibiotic resistance genes in an anaerobic membrane bioreactor treating primary clarifier effluent at 20oC. *Environmental Science: Water Research & Technology*, 4, 1783–1793. https://doi.org/10.1039/C8EW00270C

Karkman, A., Johnson, T.A., Lyra, C., Stedtfeld, R.D., Tamminen, M., Tiedje, J.M., & Virta, M. (2016). High-throughput quantification of antibiotic resistance genes from an urban wastewater treatment plant. *FEMS Microbiology Ecology*, 92(3):fiw014. https://doi.org/10.1093/femsec/fiw014

Khoshmardan, H.E., & Khoshmardan, M.E. (2021). A review of the application of nanotechnology on the removal of antibiotic-resistant bacteria and antibiotic resistance genes from discharged municipal effluents and conservation of water resources. *Fourth International Conference on Interdisciplinary Studies in Nanotechnology*, 1:01–17.

Kidd, T. J., Mills, G., Sá-Pessoa, J., Dumigan, A., Frank, C. G., Insua, J. L., ... & Bengoechea, J. A. (2017). A Klebsiella pneumoniae antibiotic resistance mechanism that subdues host defences and promotes virulence. *EMBO molecular medicine*, 9(4), 430–447.

Kourtesi, C., Ball, A.R., Huang, Y.Y., Jachak, S.M., Vera, D.M.A., Khondkar, P., Gibbons, S., Hamblin, M.R., & Tegos, G.P. (2013). Microbial efflux systems and inhibitors: approaches to drug discovery and the challenge of clinical implementation. *The Open Microbiology Journal*, 7:34–52. https://doi.org/10.2174/1874285801307010034

Kumar, A., & Pal, D. (2017). Antibiotic resistance and wastewater: Correlation, impact and critical human health challenges. *Journal of Environmental Chemical Engineering*, 6(1):52–58. https://doi.org/10.1016/j.jece.2017.11.059

Lee, C., Lee, J.H., Park, M., Park, K.S., & Bae, I.K. (2017). Biology of *Acinetobacter baumannii* : Pathogenesis, antibiotic resistance mechanisms, and prospective treatment options. *Frontiers in cellular and Infection Microbiology*, 7(55). https://doi.org/10.3389/fcimb.2017.00055

Li, D., Yu, P., Zhou, X., Kim, J.H., Zhang, Y., & Alvarez, P.J.J. (2020). Hierarchical $Bi_2O_2CO_3$ wrapped with modified graphene oxide for adsorption-enhanced photocatalytic inactivation of antibiotic resistant bacteria and resistance genes. *Water Research*, 184:116157. https://doi.org/10.1016/j.watres.2020.116157

Limayem, A., Wasson, S., Mehta, M., Pokhrel, A.R., & Patil, S. (2019). High-throughput detection of bacterial community and its drug-resistance profiling from local reclaimed wastewater plants. *Frontiers in cellular and Infection Microbiology,* 9:303. https://doi.org/10.3389/fcimb.2019.00303

Lin, M.F., & Lan, C.Y. (2014). Antimicrobial resistance in *Acinetobacter baumannii*: From bench to bedside. *World Journal of Clinical Cases,* 2:787–814. https://doi.org/10.12998/wjcc.v2.i12.787

Liu, M., Zhang, Y., Yang, M., Tian, Z., Ren, L., & Zhang, S., (2012). Abundance and distribution of tetracycline resistance genes and mobile elements in an oxytetracycline production wastewater treatment system. *Environmental Science & Technology*, 46:7551–7557. https://doi.org/10.1021/es301145m

Luczkiewicz, A., Kotlarska, E., & Artichowicz, W. (2015). Antimicrobial resistance of *Pseudomonas* spp. isolated from wastewater and wastewater-impacted marine coastal zone. *Environmental Science and Pollution Research.* 22:19823–19834. https://doi.org/10.1007/s11356-015-5098-y

Lupo, A., Coyne, S., & Berendonk, T.U. (2012). Origin and evolution of antibiotic resistance: The common mechanisms of emergence and spread in water bodies. *Frontiers in Microbiology*, 3:1–13. https://doi.org/10.3389/fmicb.2012.00018

McEwen, S.A., & Collignon, P.J. (2018). Antimicrobial resistance: A one health perspective. *Microbiology Spectrum*, 6(2). https://doi.org/10.1128/microbiolspec.ARBA-0009-2017

McLellan, S.L., Huse, S.M., Mueller-Spitz, S.R., Andreishcheva, E.N., & Sogin, M.L. (2010). Diversity and population structure of sewage-derived microorganisms in wastewater treatment plant influent. *Environmental Microbiology*, 12:378–392 https://doi.org/10.1111/j.1462-2920.2009.02075.x

Mukherjee, M., Laird, E., Gentry, T.J., Brooks, J.P., & Karthikeyan, R. (2021). Increased antimicrobial and multidrug resistance downstream of wastewater treatment plants in an urban watershed. *Frontiers in Microbiology*, 12:657353. https://doi.org/10.3389/fmicb.2021.657353

Munita, J.M., Arias, C.A., Unit, A.R., & Santiago, A.De. (2016). Mechanisms of antibiotic resistance. *Microbiology Spectrum,* 4(2):1–37. https://doi.org/10.1128/microbiolspec.VMBF-0016-2015

Nagulapally, S.R., Ahmad, A., Henry, A., Marchin, G.L., Zurek, L., & Bhandari, A. (2009). Occurrence of ciprofloxacin-, trimethoprim-sulfamethoxazole-, and vancomycin-resistant bacteria in a municipal wastewater treatment plant. *Water Environment Research,* 81(1):82–90. https://doi.org/10.2175/106143008x304596

Naquin, A., Shrestha, A., Sherpa, M., Nathaniel, R., & Boopathy, R. (2015). Presence of antibiotic resistance genes in a sewage treatment plant in Thibodaux , Louisiana , USA. *Bioresources Technology,* 188:79–83. https://doi.org/10.1016/j.biortech.2015.01.052m

Nikaido, H. (2003). Molecular basis of bacterial outer membrane permeability revisited. *Microbiology and Molecular Biology Reviews,* 67:593–656. https://doi.org/10.1128/mmbr.67.4.593-656.2003

Nirwati, H., Sinanjung, K., Fahrunissa, F., Wijaya, F., Napitupulu, S., Hati, V.P., Hakim, M.S., Meliala, A., Aman, A.T., & Nuryastuti, T. (2019). Biofilm formation and antibiotic resistance of *Klebsiella pneumoniae* isolated from clinical samples in a tertiary care hospital, Klaten, Indonesia. *BMC Proceedings,* 13(11):1–8. https://doi.org/10.1186/s12919-019-0176-7

Palleroni, N.J. (2003). Prokaryote taxonomy of the 20th century and the impact of studies on the genus *Pseudomonas*: a personal view. *Microbiology,* 149:1–7. https://doi.org/10.1099/mic.0.25952-0

Pantosti, A., & Sanchini, A. (2007). Mechanisms of antibiotic resistance in *Staphylococcus aureus. Future Microbiology,* 2:323–334. https://doi.org/10.2217/17460913.2.3.323

Peterson, E., & Kaur, P. (2018). Antibiotic resistance mechanisms in bacteria: Relationships between resistance determinants of antibiotic producers , environmental bacteria , and clinical pathogens. *Frontiers in Microbiology,* 9:2928, 1–21. https://doi.org/10.3389/fmicb.2018.02928

Piotrowska, M., & Popowska, M. (2015). Insight into the mobilome of Aeromonas strains. *Frontiers in Microbiology,* 6:494. https://doi.org/10.3389/fmicb.2015.00494

Piotrowska, M., Przygodzińska, D., Matyjewicz, K., & Popowska, M. (2017). Occurrence and variety of β-Lactamase genes among *Aeromonas* spp. isolated from urban wastewater treatment plant. *Frontiers in Microbiology,* 8:863. https://doi.org/10.3389/fmicb.2017.00863

Randall, C.P., Mariner, K.R., Chopra, I., & O'Neill, A.J. (2013). The target of daptomycin is absent form *Escherichia coli* and other gram-negative pathogens. *Antimicrobial Agents and Chemotherapy,* 57(1):637–639. https://doi.org/10.1128/aac.02005-12

Reygaert, W.C. (2018). An overview of the antimicrobial resistance mechanisms of bacteria. *AIMS Microbiology,* 4(3):482–501. https://doi.org/10.3934/microbiol.2018.3.482

Rizzo, L., Manaia, C., Merlin, C., Schwartz, T., Dagot, C., Ploy, M.C., Michael, I., & Fatta-Kassinos, D. (2013). Urban wastewater treatment plants as hotspots for antibiotic resistant bacteria and genes spread in to the environment: A review. *Science of The Total Environment,* 447:345–360.

Roberts, M.C. (2008). Update on macrolide-lincosamide-streptogramin, ketolide, and oxazolidinone resistance genes. *FEMS Microbiology Letters,* 282(2):147–159. https://doi.org/10.1111/j.1574-6968.2008.01145.x

Ruppé, É., Woerther, P.L., & Barbier, F. (2015). Mechanisms of antimicrobial resistance in Gram-negative *Bacilli. Annals of Intensive Care,* 5:21 https://doi.org/10.1186/s13613-015-0061-0

Sanderson, H., Ortega-polo, R., Mcdermott, K., Zaheer, R., Brown, R.S., Majury, A., Mcallister, T.A., & Liss, S.N. (2019). Quantification and multidrug resistance profiles of vancomycin-resistant *Enterococci* isolated from two wastewater treatment plants in the same municipality. *Microorganisms,* 7(12):626. https://doi.org/10.3390/microorganisms7120626

Schell, C.M., Tedim, A.P., Rodr, M., Sparo, M.D., Lissarrague, S., Basualdo, J.A., & Coque, T.M. (2020). Detection of β-Lactamase-producing *Enterococcus faecalis* and Vancomycin-resistant *Enterococcus faecium* isolates in human invasive infections in the public hospital of Tandil, Argentina. *Pathogens,* 9(2):142. https://doi.org/10.3390/pathogens9020142

Seiler, C., & Berendonk, T.U. (2012). Heavy metal driven co-selection of antibiotic resistance in soil and water bodies impacted by agriculture and aquaculture. *Frontiers in Microbiology,* 3:1–10. https://doi.org/10.3389/fmicb.2012.00399

Sharma, V.K., Johnson, N., Cizmas, L., McDonald, T.J., & Kim, H. (2016). A review of the influence of treatment strategies on antibiotic resistant bacteria and antibiotic resistance genes. *Chemosphere,* 150:702–714. https://doi.org/10.1016/j.chemosphere.2015.12.084

Stapleton, A.E., Wagenlehner, F.M.E., Mulgirigamam, A., & Twynholm, M. (2020). *Escherichia coli* resistance to fluoroquinolones in community-acquired uncomplicated urinary tract infection in women: A systematic review. *Antimicrobial Agents and Chemotherapy,* 64(10):e00862-20. https://doi.org/10.1128/AAC.00862-20

Sun, J., Deng, Z., & Yan, A. (2014). Bacterial multidrug efflux pumps: Mechanisms, physiology and pharmacological exploitations. *Biochemical and Biophysical Research Communications,* 453(2):254–267. https://doi.org/10.1016/j.bbrc.2014.05.090

Tesfaye, H., Alemayehu, H., Desta, A.F., & Eguale, T. (2019). Antimicrobial susceptibility profile of selected Enterobacteriaceae in wastewater samples from health facilities, abattoir, downstream rivers and a WWTP in Addis Adaba, Ethiopia. *Antimicrobial Resistance & Infection Control*, 9(8):134. https://doi.org/10.1186/s13756-019-0588-1

Teshome, A., Alemayehu, T., Deriba, W., & Ayele, Y. (2020). Antibiotic resistance profile of bacteria isolated from wastewater systems in Eastern Ethiopia. *Journal of Environmental and Public Health,* 2796365. https://doi.org/10.1155/2020/2796365

Thiolas, A., Bornet, C., Davin-Régli, A., Pagès, J-M., & Bollet, C. (2004). Resistance to imipenem, cefepime, and cefpirome associated with mutation in Omp36 osmoporin of *Enterobacter Aerogenes. Biochemical and Biophysical Research Communications,* 317:851–856. https://doi.org/10.1016/j.bbrc.2004.03.130

To, L., & Editor, T.H.E. (2013). Antimicrobial resistance profile of *Aeromonas* species isolated from Libya. *Libyan Journal of Medicine*, 8:21320. https://doi.org/10.3402/ljm.v8i0.21320.

Tobin, K.M., McGrath, J.W., Mullan, A., Quinn, J.P., & O'Connor, K.E. (2007). Polyphosphate accumulation by *Pseudomonas putida* CA-3 and other medium-chain-length polyhydroxyalkanoate-accumulating bacteria under aerobic growth conditions. *Applied and Environmental Microbiology,* 73:1383–1387. https://doi.org/10.1128/AEM.02007-06

Ubukata, K., Konno, M., & Fujii, R. (1975). Transduction of drug resistance to tetracycline, chloramphenicol, macrolides, lincomycin and clindamycin with phages induced from *Streptococcus pyogenes. The Journal of Antibiotics*, 28(9):681–688. https://doi.org/10.7164/antibiotics.28.681

Ugboko, H., & De, N. (2014). Mechanisms of Antibiotic resistance in *Salmonella typhi. International Journal of Current Microbiology and Applied Sciences,* 3(12):461–476.

Van Delden, C. (2004). Virulence factors in *Pseudomonas aeruginosa* . In: Ramos, JL. (eds.) *Pseudomonas: Virulence and Gene Regulation.* Springer, Boston, MA. https://doi.org/10.1007/978-1-4419-9084-6_1

Vaz-moreira, I., Harnisz, M., Abreu-silva, J., Rolbiecki, D., Korzeniewska, E., Luczkiewicz, A., Manaia, C.M., Plaza, G., Harnisz, M., & Abreu-silva, J. (2022). Technology Antibiotic resistance in wastewater, does the context matter? Poland and Portugal as a case study. *Critical Reviews in Environmental Science and Technology*, 52(23):4194–4216. https://doi.org/10.1080/10643389.2021.2000828

Verburg, I., Garc, S., Hern, L., Waar, K., Friedrich, A.W., & Schmitt, H. (2019). Abundance and antimicrobial resistance of three bacterial species along a complete wastewater pathway. *Microorganisms*, 7(9):312. https://doi.org/10.3390/microorganisms7090312

Vikesland, P.J., Pruden, A., Alvarez, P.J.J., Aga, D., Bu, H., Li, X., Manaia, C.M., Nambi, I., Wigginton, K., & Zhang, T. (2017). Toward a comprehensive strategy to mitigate dissemination of environmental sources of antibiotic resistance. *Environmental Science & Technology*, 51(22):13061–13069. https://doi.org/10.1021/acs.est.7b03623

Voolaid, V., Donner, E., Vasileiadis, S., & Berendonk, T.U. (2018). Bacterial diversity and antibiotic resistance genes in wastewater treatment plant influents and effluents. In Keen, P. & Fugère, R. (eds.) *Antimicrobial Resistance in Wastewater Treatment Processes,* pp. 157–178. https://doi.org/10.1002/9781119192428.ch9

Wang, H., Wang, J., Li, S., Ding, G., Wang, K., Zhuang, T., Huang, X., & Wang, X. (2020). Synergistic effect of UV/chlorine in bacterial inactivation, resistance gene removal, and gene conjugative transfer blocking. *Water Research*, 185:116290. https://doi.org/10.1016/j.watres.2020.116290

WHO (2023), Antimicrobial resistance. https://www.who.int/news-room/fact-sheets/detail/antimicrobial-resistance

Wieczorek, K., Kania, I., & Osek, J. (2013). Prevalence and antimicrobial resistance of Campylobacter spp. isolated from poultry carcasses in Poland. *Journal of food protection*, 76(8), 1451–1455.

Xue, G., Jiang, M., Chen, H., Sun, M., Liu, Y., Li, X., & Gao, P. (2019). Critical review of ARGs reduction behavior in various sludge and sewage treatment processes in wastewater treatment plants. *Critical Reviews in Environmental Science and Technology,* 49(18):1623–1674. https://doi.org/10.1080/10643389.2019.1579629

Yanagimoto, K., Yamagami, T., Uematsu, K., & Haramoto, E. (2020). Characterization of *salmonella* isolates from wastewater treatment plant influents to estimate unreported cases and infection sources of salmonellosis. *Pathogens,* 9(1):52; 5–8. https://doi.org/10.3390/pathogens9010052

Yang, S.J., Kreiswirth, B.N., Sakoulas, G., Yeaman, M,R., Xiong, Y.Q., Sawa, A., & Bayer, A.S. (2009). Enhanced expression of *dltABCD* is associated with development of daptomycin nonsusceptibility in a clinical endocarditis isolate of *Staphylococcus aureus*. *The Journal of Infectious Diseases*, 200(12):1916–1920. https://doi.org/10.1086/648473

Yang, Y., Li, B., & Zou, S. (2014). Fate of antibiotic resistance genes in sewage treatment plant revealed by metagenomic approach. *Water Research*, 62(1):97–106. https://doi.org/10.1016/j.watres.2014.05.019

Ye, L., & Zhang, T. (2013). Bacterial communities in different sections of a municipal wastewater treatment plant revealed by 16S rDNA 454 pyrosequencing. *Applied Microbiology and Biotechnology*, 97:2681–2690. https://doi.org/10.1007/s00253-012-4082-4

Zhang, T., Zhang, M., Zhang, X., & Fang, H. (2009). Tetracycline resistance genes and tetracycline resistant lactose-fermenting Enterobacteriaceae in activated sludge of sewage treatment plants. *Environmental Science & Technology*, 43(10):3455–3460. https://doi.org/10.1021/es803309m

Zhang, Y., Marrs, C.F., Simon, C., & Xi, C. (2009). Wastewater treatment contributes to selective increase of antibiotic resistance among *Acinetobacter* spp. *Science of The Total Environment*, 407(12):3702–3706. https://doi.org/10.1016/j.scitotenv.2009.02.013

Zielin´ski, W., Korzeniewska, E., Harnisz, M., Hubeny, J., Buta, M., & Rolbiecki, D. (2020). The prevalence of drug-resistant and virulent *Staphylococcus* spp. in a municipal wastewater treatment plant and their spread in the environment. *Environment International*, 143:105914. https://doi.org/10.1016/j.envint.2020.105914

10 A Modern Focus on Foodborne Microbes

Antimicrobial Properties of Food Nano-Packaging

P.F. Steffi, B. Thirumalaiyammal, N. Sathammai Priya, and Thenmozhi Manivel

10.1 INTRODUCTION

Bacteria, parasites, and viruses can cause foodborne diseases, which are classified as foodborne infections and foodborne intoxication. The transmission of foodborne infections occurs when a pathogen establishes itself in the human host and multiplies. Foodborne intoxication occurs when a pathogen produces toxins in food that are ingested by humans (Bintsis T, 2017). In 2015, the World Health Organization reported that foodborne diarrhoeal diseases caused approximately 550 million cases and 230,000 deaths worldwide. Foodborne disease outbreaks are on the rise, posing a serious threat to the health of consumers. Increasing awareness of such concerns has prompted researchers to develop innovative and potent foodborne disease prevention agents. Such endeavours could potentially have a future impact in critical situations, including pandemic diseases like influenza viruses, severe acute respiratory syndrome (SARS), and its more recent variant, SARS-CoV-2 (coronavirus disease 2019, COVID-19). The overall rate of foodborne disease outbreaks leads to great concern about contagious food products; thus, significant emphasis is being placed on the development of new and effective agents against food pathogenic microorganisms. These efforts might have a possible future impact in other critical conditions, similar to epidemic conditions like influenza viruses, SARS, and its newer form, SARS-CoV-2 (Coronavirus disease 2019, COVID-19).

Food packaging is usually used to protect the quality of food and make it last longer. The right packaging can help protect food from microbial damage or other environmental contamination, but if it's not of good quality, it can lead to more meals going bad and more foodborne illnesses. Traditional meal applications are just passive barriers that can only postpone the bad effects of environmental contamination (WHO, 2015). The goal of classic packaging is to be as inert as possible in contact with food (Brody A et al., 2008). The development of novel packaging is becoming a major area of research in the present day. This new type of packaging can be used to reduce the risk of pathogen contamination and extend the shelf life of food. The application of nanotechnology provides novel opportunities to investigate the bactericidal potential of nano-based materials with high bioactivity. Nano is the study and utilization of materials and structures with a resolution of 1 to 100 nanometres (9–10 nanometres or one-billionth of a nanometre) (Biji K et al., 2015). Today, innovation in non-traditional packaging is a key area of research. The new generation of food packaging can reduce the risk of pathogen contamination and extend the shelf life of food. Nanotechnology introduces a new food packaging technology for the food industry. The term "nano-packaging" is derived from the combination of the terms "nanotechnology" and "food packaging". demonstrating

DOI: 10.1201/9781003268925-10

a direct application of nanoscience in food science. Nanoscience creates and uses materials with new physio-chemical properties that provide many new opportunities for food industries (Sujithra S et al., 2019). These opportunities include more powerful food colouring, flavouring, nutritional additives, and antibacterial additives for food packaging.

The integration of nanotechnology into the food industry has paved the way for innovative solutions aimed at enhancing food safety and quality. One of the significant advancements in this domain is the development of food nano-packaging with inherent antimicrobial properties. Foodborne illnesses caused by microbial contamination remain a persistent global health concern, leading to significant economic losses and public health implications. Traditional food packaging methods provide passive protection against contamination, but the emergence of nanomaterials has opened new avenues for actively inhibiting the growth of microorganisms. Antimicrobial nanomaterials, such as silver nanoparticles, zinc oxide nanoparticles, and chitosan nanoparticles, offer potent and broad-spectrum antimicrobial effects. When integrated into food packaging, these nanomaterials can significantly extend the shelf life of perishable food products and enhance food safety by preventing the growth of spoilage microorganisms and foodborne pathogens. This paper explores the antimicrobial properties of food nano-packaging, highlighting the potential benefits it offers in revolutionizing food preservation practices and contributing to a safer and more sustainable food supply chain.

Antimicrobial nanomaterials are particularly interesting due to their barrier properties and desired structural integrity, which reduces the spoilage and the growth of pathogenic microorganisms in antimicrobial films. Nanomaterials can be used in antimicrobial films as growth inhibitors, killing agents or even antibiotic carriers. Several studies demonstrate the biological activity of inorganic nanoparticles with high antibacterial activity at low concentrations. Inorganic nanoparticles are the most common inorganic nanomaterials used for food packaging. Silver nanoparticles are a popular inorganic nanomaterial used for food packaging due to their excellent toxicity to a wide range of microorganisms, low volatility, and high-temperature stability (Gupta, A., et al., 2016). The most common method for the preparation of AgNPs is chemical reduction, where silver ions are reduced to silver atoms and aggregated into colloidal clusters (Sujithra, S., et al., 2019). Nanomaterials have a wide range of applications in the food nano-packaging industry due to their fundamental differences in structure and physicochemical properties. While there have been many informative studies on the practical applications of different nanomaterials in the food packaging industry, there are a limited number of studies in food microbiology that focus on the use of nano-packaging technology. While nano-packaging as antimicrobial properties, it's important to emphasize that the interactions of nanomaterials with food ingredients, and potential changes in food quality, range from sensory to safety aspects (Jafarzadeh, S., et al., 2020). This is largely dependent on the nanomaterial doses and concentration used in the fabrication of nano-packages. Optimizing the correct dose and concentration that allow controlled release of the nanoparticle can maintain antimicrobial properties without altering the food quality and safety (Kavakebi E. et al., 2021).

10.2 FOODBORNE PATHOGENS IN FOOD PACKAGING

10.2.1 Importance of Food Packaging in Food Safety

Food packaging plays a pivotal role in ensuring food safety by safeguarding the quality and integrity of food products throughout the supply chain. It serves as a barrier against physical, chemical, and microbial contaminants, effectively preventing potential hazards that could lead to foodborne illnesses. Proper packaging helps maintain the freshness of perishable items, extends their shelf life, and protects against environmental factors such as light, oxygen, and moisture, which can compromise food quality. Additionally, innovative packaging technologies, such as active and intelligent packaging, antimicrobial coatings, and nanotechnology applications, have emerged as powerful tools to enhance food safety. These advancements provide real-time monitoring, traceability, and antimicrobial properties, further contributing to the prevention of contamination and foodborne diseases. Moreover, food packaging ensures clear labelling and allergen information, enabling consumers to make informed choices and avoid potential allergens or harmful ingredients. By complying with global food packaging regulations and standards, manufacturers can maintain consistency and safety in packaging materials, thereby mitigating potential risks to consumer health. Overall, the strategic use of food packaging is fundamental in protecting public health and instilling confidence in the safety and reliability of the food supply (WHO 2015). The importance of food packaging in food safety can be highlighted through the following key points:

10.2.1.1 Physical Protection

Packaging provides a physical barrier that shields food from external elements, such as dust, dirt, and pests. It prevents physical damage during handling and transportation, reducing the risk of contamination and maintaining the integrity of the food product.

10.2.1.2 Barrier to Chemical Contaminants

Food packaging acts as a barrier to prevent the migration of harmful chemicals or substances from the environment into the food. It helps to avoid potential contamination from packaging materials, ink, or any other external sources.

10.2.1.3 Microbial Safety

Properly designed food packaging can significantly reduce the risk of microbial contamination. It prevents the entry of microorganisms like bacteria, yeasts, and moulds that can cause foodborne illnesses and spoilage. Additionally, active and intelligent packaging technologies, including antimicrobial nano-packaging, can actively inhibit the growth of pathogens and extend the shelf life of food products.

10.2.1.4 Product Quality Maintenance

Food packaging helps maintain the quality and freshness of food by protecting it from exposure to air, light, and moisture. This is especially important for perishable items like fruits, vegetables, and meat, as well as sensitive products like dairy and ready-to-eat meals.

10.2.1.5 Information and Communication

Packaging provides essential information to consumers, such as ingredient lists, nutrition facts, allergen warnings, and expiry dates. Clear labelling and proper packaging facilitate informed decision-making and reduce the risk of food-related health issues.

10.2.1.6 Convenience and Portability

Packaging contributes to the convenience and portability of food products, enabling consumers to carry and consume them safely and easily. Properly sealed and portion-sized packaging reduces the risk of cross-contamination and overconsumption.

10.2.1.7 Food Waste Reduction

Adequate packaging can reduce food waste by preserving the freshness and quality of food products. It extends shelf life, allowing more time for products to be consumed before spoilage occurs, thereby minimizing food waste. Food packaging regulations, standards, and guidelines are in place to ensure the safety and suitability of packaging materials used for food contact.

10.2.2 Understanding Foodborne Pathogens and Their Impact on Public Health

Foodborne pathogens are microorganisms that can contaminate food and cause illnesses when ingested by humans. They include bacteria, viruses, parasites, and fungi that can be present in raw or undercooked foods, contaminated water, and cross-contaminated surfaces. These pathogens pose a significant public health concern worldwide, leading to millions of cases of foodborne illnesses and even fatalities every year. The impact of foodborne pathogens on public health can range from mild gastrointestinal discomfort to severe and life-threatening conditions. Symptoms may include nausea, vomiting, diarrhoea, abdominal pain, fever, and in more severe cases, organ failure. Vulnerable populations such as children, the elderly, pregnant women, and individuals with compromised immune systems are at higher risk of developing severe complications from foodborne infections. Understanding the sources of contamination, proper food handling and preparation, and implementing effective food safety measures are crucial to preventing the spread of foodborne pathogens and protecting public health (FAO, 2011).

Understanding foodborne pathogens and their impact on public health is essential for implementing effective preventive measures and ensuring food safety. Here are key points to consider:

1. **Common foodborne pathogens:** Several types of bacteria, viruses, parasites, and toxins can contaminate food and cause foodborne illnesses. Common foodborne pathogens include *Salmonella*, *Campylobacter*, *Escherichia coli* (*E. coli*), *Listeria* monocytogenes, norovirus, hepatitis A virus, and various types of foodborne toxins produced by certain moulds or bacteria (WHO (2015).
2. **Routes of contamination:** Foodborne pathogens can contaminate food at various points along the food supply chain, including during production, processing, distribution, and preparation. Contamination can occur through contaminated water, soil, equipment, and contact with infected animals or humans.
3. **Symptoms and impact on health:** Foodborne illnesses can range from mild gastrointestinal discomfort to severe and life-threatening conditions. Symptoms often include nausea, vomiting, diarrhoea, abdominal pain, fever, and in some cases, neurological or systemic complications. Vulnerable populations, such as the elderly, young children, pregnant women, and immunocompromised individuals, are at a higher risk of severe outcomes.
4. **Burden of foodborne diseases:** Foodborne illnesses have a considerable impact on public health. They result in millions of cases, hospitalizations, and deaths each year globally. Besides the human toll, foodborne diseases impose a substantial economic burden through medical costs, loss of productivity, and food recalls.
5. **Outbreaks and surveillance:** Foodborne disease outbreaks can occur when a common source of contaminated food affects multiple individuals. Timely and accurate surveillance systems are crucial for identifying outbreaks, tracking foodborne pathogens, and implementing appropriate control measures.
6. **Preventive measures:** To reduce the impact of foodborne pathogens on public health, various preventive measures are essential. These include good agricultural practices,

proper food handling, processing, and storage, hygienic practices during food preparation, adequate sanitation, and effective monitoring and regulation of the food industry.

7. **Risk communication:** Effective communication between public health authorities, the food industry, and the public is vital for raising awareness of foodborne pathogens, promoting safe food practices, and responding to outbreaks promptly.

10.2.3 Role of Food Packaging in Preventing Pathogen Contamination

The role of food packaging in preventing pathogen contamination is paramount in ensuring food safety and protecting public health. Packaging acts as a physical barrier that shields food products from external sources of contamination, including harmful pathogens. It serves as a protective layer, preventing direct contact between the food and potential sources of contamination, such as unclean surfaces, handling practices, and environmental contaminants. Properly sealed and secure packaging prevents the entry of bacteria, viruses, and other pathogens that could otherwise contaminate the food during storage, transportation, and handling. Furthermore, food packaging helps maintain the integrity of the product by preventing leaks, spills, and exposure to unsanitary conditions, which can lead to microbial growth and contamination. Packaging also extends the shelf life of perishable items, reducing the risk of pathogen growth during storage and distribution. In addition to physical barriers, innovative packaging technologies, such as antimicrobial coatings and active packaging, provide an extra layer of protection by actively inhibiting the growth of pathogens and spoilage microorganisms. By employing proper food packaging practices, the risk of pathogen contamination and subsequent foodborne illnesses can be significantly reduced, contributing to improved food safety and public health. Food manufacturers and distributors must adhere to strict packaging guidelines and standards to ensure that their products remain safe for consumption throughout the supply chain (Friedman M et al., 2002). Here are some ways in which food packaging prevents pathogen contamination:

1. **Physical barrier:** Packaging materials create a physical barrier that protects food from direct contact with potential sources of contamination, such as dust, dirt, insects, and human handling. It prevents external pathogens from coming into contact with the food, reducing the risk of contamination.
2. **Tamper-evident packaging:** Tamper-evident packaging helps to prevent intentional contamination or adulteration of food products. If the packaging is compromised, it becomes evident, allowing consumers to identify potential tampering and avoid consuming the product.
3. **Protection from cross-contamination:** Packaging helps to separate different food items and prevent cross-contamination. For instance, raw meat is often packaged separately from ready-to-eat foods to avoid the transfer of harmful pathogens from raw to cooked food.
4. **Airtight seals:** Airtight seals in packaging prevent the entry of air, moisture, and other external contaminants, creating an inhospitable environment for microbial growth and spoilage.
5. **Barrier to moisture:** Moisture can facilitate the growth of bacteria, moulds, and other microorganisms. Packaging materials with moisture barriers help to maintain the dryness of food products, reducing the risk of microbial contamination.
6. **Antimicrobial packaging:** Advanced packaging technologies incorporate antimicrobial agents or nanoparticles that actively inhibit the growth of pathogens on the food's surface. These antimicrobial packaging materials can help to further enhance food safety.
7. **Time–temperature indicators:** Some packaging includes time–temperature indicators that change colour or provide visual cues if the product has been exposed to temperature abuse during storage or transportation. This helps consumers and retailers identify potentially unsafe food.

8. **Controlled atmosphere packaging:** Modified atmosphere or controlled atmosphere packaging adjusts the composition of gases surrounding the food product, slowing down microbial growth and extending shelf life.

10.2.4 Common Foodborne Pathogens in Packaging

Common foodborne pathogens can inadvertently contaminate food packaging during various stages of the food production and distribution process, posing a potential risk to consumers. *Salmonella* spp., *Escherichia coli* (*E. coli*), *Listeria monocytogenes*, *Campylobacter* spp., and *Staphylococcus aureus* are among the most frequently encountered pathogens associated with food packaging. These pathogens can originate from contaminated raw ingredients, cross-contamination during processing, inadequate sanitation practices, or improper handling during packaging operations. Once introduced into the packaging, these pathogens can survive and potentially multiply, posing a threat to food safety. Implementing rigorous hygiene practices, adhering to good manufacturing processes, and using antimicrobial packaging materials are crucial strategies to minimize the risk of pathogen contamination in food packaging and ensure the safety of food products (Li, W. R., et al., 2011) (Figure 10.1). Some of the most prevalent foodborne pathogens found in packaging include:

1. ***Salmonella*:** *Salmonella* is a widespread bacterial pathogen that can contaminate food products, including packaging materials. It can survive on surfaces and has been known to cause outbreaks when it comes into contact with packaging during food processing or handling.
2. ***Listeria monocytogenes*:** *Listeria* is a dangerous pathogen that can survive and grow at refrigeration temperatures. It has been linked to outbreaks associated with ready-to-eat foods and has been found on the surfaces of packaging materials.

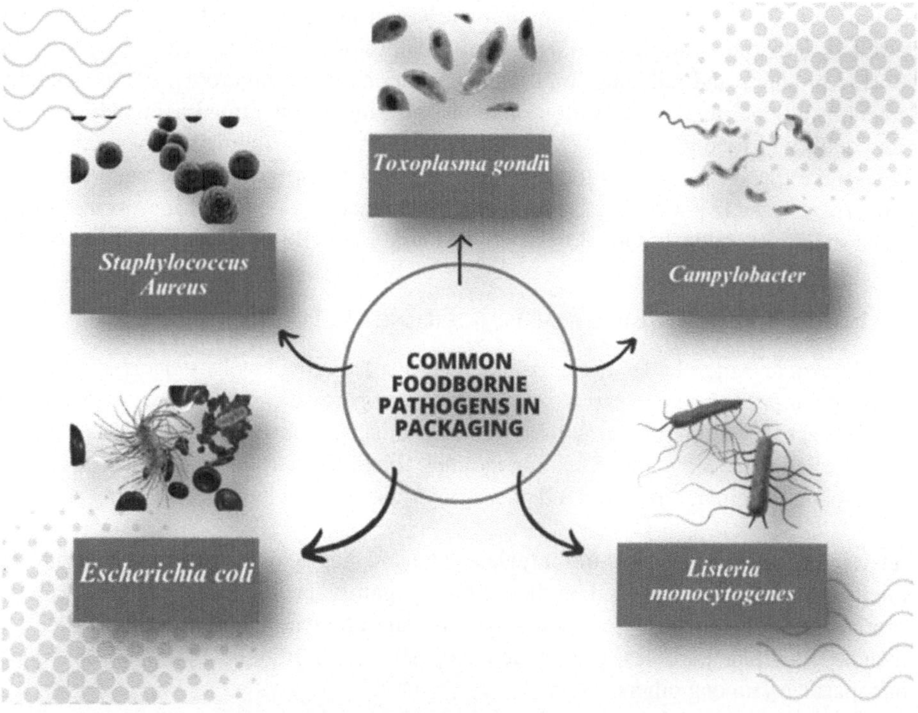

FIGURE 10.1 Common foodborne pathogens in packaging.

3. ***Escherichia coli* (*E. coli*):** Certain strains *of E. coli*, such as *E. coli* O157:H7, are pathogenic and can cause severe illness. Contamination of packaging materials with *E. coli* can occur through contact with contaminated hands, equipment, or water sources.
4. ***Campylobacter*:** *Campylobacter* is a common bacterial pathogen associated with raw poultry products. It can potentially contaminate packaging materials during the packaging process.
5. **Norovirus:** Norovirus is a highly contagious virus that can cause gastroenteritis outbreaks. While it is primarily transmitted through contaminated food, it can also be transferred to food packaging surfaces during handling.
6. **Hepatitis A virus:** Hepatitis A virus is another viral pathogen that can contaminate food packaging, particularly through contact with infected handlers.
7. ***Staphylococcus aureus*:** *Staphylococcus aureus* is a common bacteria found on human skin and can contaminate food packaging if handled by infected individuals.

It is important to note that the risk of foodborne pathogens in packaging can be mitigated through proper hygiene, sanitation, and good manufacturing practices in the food industry. Regular cleaning and disinfection of equipment, surfaces, and packaging materials are essential for preventing cross-contamination.

10.3 ANTIMICROBIAL NANO-PACKAGING

Antimicrobial nano-packaging is an innovative approach that leverages nanotechnology to incorporate antimicrobial properties into food packaging materials. Nanoparticles, such as silver, zinc oxide, and copper, are embedded or coated onto the packaging surface, imparting antimicrobial activity against a wide range of pathogens. These nanoparticles release antimicrobial agents in a controlled manner, effectively inhibiting the growth and survival of bacteria, viruses, and other microorganisms that could contaminate food products. The use of antimicrobial nano-packaging offers several advantages in enhancing food safety. First, it provides an additional layer of protection, minimizing the risk of pathogen contamination during storage and transportation. Second, the controlled release of antimicrobial agents ensures prolonged efficacy, extending the shelf life of perishable products and reducing food waste. Additionally, antimicrobial nano-packaging can complement traditional food preservation methods, offering an integrated approach to ensure the safety and quality of food products. While antimicrobial nano-packaging shows significant promise, it is essential to carefully assess the safety and regulatory aspects of these materials. Ensuring that nanoparticles used in packaging are safe for food contact is crucial, and appropriate testing and evaluations should be conducted to address any potential risks (Rai M et al., 2009) (Table 10.1).

10.3.1 Integration of Antimicrobial Immaterialness in Food Packaging

The integration of antimicrobial nanomaterials into food packaging involves incorporating nanoparticles or nanocomposites with inherent antimicrobial properties into packaging materials. These nanomaterials actively inhibit the growth of microorganisms, such as bacteria, yeasts, and moulds, thereby enhancing food safety and extending the shelf life of food products. The process of integrating antimicrobial nanomaterials into food packaging typically involves the following steps:

1. **Selection of antimicrobial nanomaterials:** The first step is to choose the appropriate antimicrobial nanomaterials based on their efficacy against specific foodborne pathogens and spoilage microorganisms. Common antimicrobial nanomaterials used in food packaging include silver nanoparticles, zinc oxide nanoparticles, chitosan nanoparticles, and copper nanoparticles, among others.
2. **Nanomaterial dispersion:** Once the antimicrobial nanomaterials are selected, they are dispersed uniformly into the packaging material. This can be achieved through techniques

TABLE 10.1
Nano Antimicrobial Agents in the Preservation of Food

Sr. No	Nanoparticles	Tested Food	References
1	Silver nanoparticles	Barberry	(22)
2	Silver nanoparticles and zinc oxide nanoparticles	Orange juice	(23)
3	Silver nanoparticles and titanium dioxide nanoparticles	Fresh apples, bread, carrot juice, orange juice	(24)
4	Zinc oxide nanoparticle	Meat	(25)

such as melt mixing, extrusion, or solvent casting, depending on the type of packaging material (e.g. polymer-based or biodegradable) and the desired distribution of nanoparticles.

3. **Surface coating:** In some cases, nanolayers or nanocoatings of antimicrobial nanomaterials are applied to the surface of the packaging material. These nanocoatings can provide an additional layer of protection against microbial contamination and maintain the functionality of the antimicrobial nanomaterials.
4. **Controlled release systems:** Antimicrobial nanomaterials can be incorporated into controlled release systems, such as nanoliposomes or nanoemulsions. These systems enable the sustained release of antimicrobial agents over time, providing prolonged protection against microorganisms.
5. **Real-time monitoring:** Nanosensors and indicators can be integrated into the packaging to provide real-time monitoring of critical parameters, such as temperature, gas composition, and pathogen presence. This active monitoring ensures better food safety and quality control during storage and transportation.
6. **Safety assessment:** Before using antimicrobial nanomaterials in food packaging, safety assessments must be conducted to evaluate their potential toxicity, migration into food, and impact on human health. Regulatory bodies have established guidelines for the safe use of nanomaterials in food contact applications.

The integration of antimicrobial nanomaterials into food packaging offers several advantages, including improved food safety, extended shelf life, reduced food waste, and reduced dependence on chemical preservatives. However, it is essential to ensure that the antimicrobial nanomaterials used are safe, effective, and compliant with food contact regulations to protect consumer health and maintain the integrity of the food supply chain.

10.3.2 Mechanism of Antimicrobial Nano-Packaging

Antimicrobial nano-packaging harnesses the unique properties of nanoparticles to combat microbial contamination in food packaging materials. The antimicrobial mechanisms of nanoparticles vary depending on their composition, size, and surface properties. Silver nanoparticles (AgNPs) release silver ions that interact with bacterial cell membranes, leading to membrane disruption and subsequent cell death. Similarly, copper nanoparticles (CuNPs) can penetrate bacterial cell walls, interfering with their cellular functions and causing microbial inactivation. Zinc oxide nanoparticles (ZnONPs) induce oxidative stress within microorganisms, leading to the production of reactive oxygen species (ROS) that damage cellular components and inhibit growth. Titanium dioxide nanoparticles (TiO2NPs) possess photocatalytic activity and can generate ROS when exposed to UV light, which leads to the degradation of pathogenic cells. Chitosan nanoparticles have a positive charge that allows them to interact with negatively charged bacterial cell membranes, leading to membrane permeabilization and cell lysis. By utilizing these diverse mechanisms, antimicrobial nano-packaging provides an effective and multifaceted approach to prevent pathogen contamination, enhance food safety, and extend the shelf life of food products (Carbone M et al., 2019).

10.3.3 Types of Nanomaterials Used in Food Packaging

10.3.3.1 Silver Nanoparticles (AgNPs)

Silver nanoparticles (AgNPs) have emerged as a promising antimicrobial agent in food packaging due to their potent inhibitory effects against a wide range of microorganisms. AgNPs possess unique physicochemical properties that enable them to interact with bacterial cell membranes, disrupting their integrity and causing cell death. The release of silver ions from AgNPs further enhances their antimicrobial activity by interfering with the vital cellular processes of pathogens. Incorporating AgNPs into food packaging materials can effectively prevent microbial growth, contamination, and spoilage during storage and transportation, thereby extending the shelf life of perishable food products. Moreover, the use of AgNPs in food packaging aligns with the growing demand for safer and more sustainable food preservation techniques. However, safety and regulatory considerations remain essential to ensure the appropriate use of AgNPs in food contact materials, promoting consumer confidence and public health (Vimbela G. V. et al., 2017).

10.3.3.2 Zinc Oxide Nanoparticles (ZnONPs)

Zinc oxide nanoparticles (ZnONPs) have shown great potential in antimicrobial nano-packaging due to their exceptional antimicrobial properties and versatile applications. ZnONPs possess a unique ability to inhibit the growth and proliferation of a wide range of microorganisms, including bacteria, viruses, and fungi. When incorporated into food packaging materials, ZnONPs can serve as an effective barrier against pathogen contamination, preventing the transfer of harmful microorganisms to food products. The controlled release of antimicrobial agents from ZnONPs ensures sustained and prolonged protection, making them highly efficient in preserving the quality and safety of food items during storage and transportation. Research has demonstrated the efficacy of ZnONPs in reducing the populations of foodborne pathogens such as *Escherichia coli*, *Salmonella* spp., and *Staphylococcus aureus*. Additionally, ZnONPs have shown promise in controlling spoilage microorganisms, extending the shelf life of perishable foods, and minimizing food waste. As a result, antimicrobial nano-packaging utilizing ZnONPs has the potential to significantly enhance food safety, reduce the risk of foodborne illnesses, and improve overall public health (Bai H et al., 2016).

10.3.3.3 Copper Nanoparticles (CuNPs)

Copper nanoparticles (CuNPs) have emerged as a promising antimicrobial agent in the context of antimicrobial food packaging. CuNPs possess inherent antimicrobial properties, making them effective against a wide range of microorganisms, including bacteria, viruses, and fungi. When incorporated into food packaging materials, CuNPs create an additional barrier against pathogen contamination, helping to prevent the growth and transmission of harmful microorganisms to food products. The antimicrobial action of CuNPs is attributed to their ability to penetrate bacterial cell walls and interfere with essential cellular functions, leading to microbial inactivation. Studies have demonstrated the effectiveness of CuNPs in reducing the populations of foodborne pathogens such as *Escherichia coli*, *Listeria monocytogenes*, and *Staphylococcus aureus*. Moreover, CuNPs have exhibited excellent inhibitory effects on spoilage microorganisms, contributing to the extension of the shelf life of perishable foods and improving overall food quality. The use of CuNPs in antimicrobial food packaging holds promise in enhancing food safety, reducing the risk of foodborne illnesses, and ensuring the delivery of safe and wholesome food products to consumers (Kumar P et al., 2019).

10.3.3.4 Titanium Dioxide Nanoparticles (TiO_2NPs)

Titanium dioxide nanoparticles (TiO_2NPs) have gained significant attention in the realm of antimicrobial nano-packaging due to their remarkable antimicrobial properties and wide-ranging applications. TiO_2NPs possess photocatalytic activity, which means that when exposed to ultraviolet (UV) light, they generate ROS such as superoxide radicals and hydroxyl radicals. These ROS possess potent antimicrobial effects, as they can cause damage to the cell membranes and biomolecules of microorganisms, leading to their inactivation. Incorporating TiO_2NPs into food packaging materials offers an innovative approach to prevent microbial contamination and enhance food safety. The photocatalytic action of TiO_2NPs provides continuous and long-lasting antimicrobial effects, making them highly effective in inhibiting the growth of bacteria, viruses, and fungi that may come into contact with the packaging surface. Moreover, TiO_2NPs can contribute to extending the shelf life of food products by controlling the proliferation of spoilage microorganisms. This can help reduce food waste and ensure the delivery of high-quality and safe food items to consumers. It is important to note that the safety and regulatory aspects of TiO_2NPs in food contact materials must be carefully considered to ensure their safe and effective use in antimicrobial nano-packaging (Durán-Lara E. F. et al., 2020).

10.3.3.5 Chitosan Nanoparticles and Other Biopolymers

Chitosan nanoparticles and other biopolymers have shown immense potential as antimicrobial nano-packaging materials due to their natural origin, biocompatibility, and inherent antimicrobial properties. Chitosan, derived from chitin found in crustacean shells, possesses a positively charged structure, enabling it to interact with negatively charged bacterial cell membranes. This interaction leads to membrane destabilization and cell lysis, effectively inhibiting the growth and survival of a wide range of microorganisms, including bacteria and fungi. In addition to chitosan, other biopolymers such as starch, cellulose, and alginate have also been explored for their antimicrobial properties in nano-packaging. These biopolymers can be processed into nanoparticles, which offer a large surface area for antimicrobial interactions and controlled release of antimicrobial agents. The utilization of chitosan nanoparticles and other biopolymer-based materials in antimicrobial nano-packaging provides a sustainable and eco-friendly approach to enhance food safety. The incorporation of these natural materials into food packaging can effectively prevent microbial contamination, extend the shelf life of perishable food products, and contribute to reducing the risk of foodborne illnesses (Arfat Y. A et al., 2021) (Table 10.2).

TABLE 10.2
Examples of Foods, and Food Packaging That Contain Nanomaterials

Sr. No	Type of Product	Product Name and Manufacturer	Nano Content
1	Beverage	Oat chocolate and oat vanilla nutritional drink mixes; toddler health	300 nm particles of iron (SunActive Fe)
2	Food additive	Bioral™ Omega-3 nanocochleates; BioDelivery Sciences Internation	Nano-cochleate as small as 50 nm
3	Food packaging	Food packaging Durethan® KU 2-2601 plastic wrapping; Bayer	Nanoparticles of silica in a polymer-based nanocomposite
4	Food packaging	PrimoMaxx, Syngenta	100 nm particle size emulsion
5	Food contact material	Nanosilver cutting board; A-Do Global	Nanoparticles of silver
6	Food contact material	Antibacterial kitchenware; Nano Care Technology/NCT	Nanoparticles of silver

10.4 COMMON NANOPARTICLES FOR LOADING ANTIMICROBIAL AGENTS IN ACTIVE PACKAGING

In active packaging, nanocarriers are used to encapsulate and deliver antimicrobial agents effectively. Nanocarriers are nano-sized structures that can protect and control the release of active compounds, enhancing their stability and efficacy. The following are some common nanocarriers for loading antimicrobial agents in active packaging.

10.4.1 Liposomes

Liposomes are spherical vesicles composed of lipid bilayers that can encapsulate hydrophilic and hydrophobic antimicrobial agents. They provide a controlled release mechanism and protect the antimicrobial compounds from degradation, increasing their effectiveness in inhibiting microbial growth.

10.4.2 Nanoemulsions

Nanoemulsions are oil-in-water or water-in-oil dispersions stabilized by surfactants. They can encapsulate hydrophobic antimicrobial compounds and maintain their solubility and stability. Nanoemulsions offer improved penetration and distribution of antimicrobial agents on the surface of food, enhancing their antimicrobial activity.

10.4.3 Nanofibres

Electrospun nanofibres can be used as carriers for antimicrobial agents. These nanofibres have a high surface area and can be functionalized to release antimicrobial compounds gradually. They are commonly used as coatings on food packaging materials to provide continuous antimicrobial protection.

10.4.4 Nanocomposites

Nanocomposites are materials formed by incorporating nanoparticles into a polymer matrix. They can load and release antimicrobial agents in a controlled manner, providing long-lasting protection against pathogens on the surface of food (Emamifar A et al., 2010).

10.4.5 Nanogels

Nanogels are three-dimensional networks of cross-linked polymers capable of absorbing and releasing water-soluble antimicrobial agents. They offer a stable and sustained release of antimicrobial compounds, making them suitable for active packaging applications.

10.4.6 Solid Lipid Nanoparticles (SLNs) and Nanostructured Lipid Carriers (NLCs)

SLNs and NLCs are lipid-based nanocarriers that can encapsulate both lipophilic and hydrophilic antimicrobial agents. They protect against degradation and provide controlled release, ensuring prolonged antimicrobial activity.

These nanocarriers offer several advantages, including improved stability, controlled release, and enhanced bioavailability of antimicrobial agents. By using nanocarriers in active packaging, the efficiency and performance of antimicrobial agents can be significantly enhanced, making them more suitable for food preservation and safety applications (Motlagh N.V. et al., 2012).

10.5 APPLICATION OF ANTIMICROBIAL NANO-PACKAGING IN FOOD PRESERVATION

The application of antimicrobial nano-packaging in food preservation holds significant promise in enhancing food safety and extending the shelf life of perishable food products. Antimicrobial nanoparticles, such as silver nanoparticles (AgNPs), zinc oxide nanoparticles (ZnONPs), and chitosan nanoparticles, can be incorporated into food packaging materials to create a protective barrier against microbial contamination. These nanoparticles release antimicrobial agents in a controlled manner, effectively inhibiting the growth and survival of bacteria, viruses, and fungi that may come into contact with the packaging surface. By preventing the proliferation of harmful microorganisms, antimicrobial nano-packaging helps reduce the risk of foodborne illnesses and spoilage, thereby minimizing food waste. The sustained release of antimicrobial agents from nanoparticles ensures long-lasting protection during storage and transportation, maintaining the freshness and quality of food products. Additionally, antimicrobial nano-packaging can complement traditional food preservation methods, providing an integrated approach to ensure the safety and quality of food items throughout the supply chain. As a result, the application of antimicrobial nano-packaging in food preservation represents a promising strategy to enhance food safety, improve food quality, and meet the increasing demand for safer and more sustainable food packaging solutions. They are given: 1. Extended Shelf Life and Quality Maintenance; 2. Pathogen Control and Contamination Prevention; 3. Preservation of Nutritional Value.

10.6 CHALLENGES AND FUTURE PERSPECTIVES

10.6.1 Challenges

1. **Safety and regulation:** The safety of using nanoparticles in food packaging is a critical concern. Regulatory agencies require comprehensive safety assessments to ensure that nanoparticles do not pose any health risks to consumers. Evaluating the potential toxicological effects and establishing appropriate safety standards are essential for gaining regulatory approval and public acceptance.
2. **Cost-effectiveness:** The production and incorporation of antimicrobial nanoparticles into food packaging materials can be expensive. Finding cost-effective methods for nanoparticle synthesis and scaling up production is crucial for the widespread adoption of antimicrobial nano-packaging in the food industry.
3. **Compatibility with food types:** Different food products have unique characteristics, and the effectiveness of antimicrobial nanoparticles may vary with different food types. Factors such as pH, temperature, and food composition can influence the antimicrobial efficacy of nanoparticles. Tailoring nano-packaging solutions to specific food applications is necessary to ensure optimal performance.
4. **Nanoparticle stability:** The stability and longevity of nanoparticles in food packaging need to be addressed. Factors such as nanoparticle leaching, migration, and agglomeration can affect their antimicrobial efficacy and safety over time.
5. **Consumer perception:** Public perception and acceptance of nanotechnology in food packaging can influence its market adoption. Transparent communication about the benefits, safety, and necessity of antimicrobial nano-packaging is crucial to address any consumer concerns.

10.6.2 Future Perspectives

1. **Advanced nanomaterials:** Ongoing research in nanotechnology will likely yield advanced nanomaterials with enhanced antimicrobial properties, improved stability, and reduced

production costs. These advancements will open up new possibilities for effective and affordable antimicrobial nano-packaging solutions.

2. **Integrating technologies:** Combining antimicrobial nanoparticles with other food preservation technologies, such as active packaging or edible coatings, can lead to synergistic effects, further enhancing the antimicrobial performance and extending the shelf life of food products.
3. **Personalized packaging:** Tailoring nano-packaging solutions to specific food products and consumer preferences can optimize food safety and quality maintenance, catering to the diverse demands of the food industry.
4. **Sustainable nanomaterials:** Developing eco-friendly and biodegradable nanoparticles will address environmental concerns and promote sustainable antimicrobial nano-packaging practices, aligning with the growing interest in eco-conscious food packaging solutions.
5. **Industry collaboration:** Collaboration between researchers, food manufacturers, and regulatory bodies is crucial for driving innovation and establishing safety standards for antimicrobial nano-packaging. Such partnerships will expedite the safe adoption of this technology in the food industry.

10.7 COMPARISON WITH TRADITIONAL PACKAGING METHODS

Antimicrobial nano-packaging and traditional packaging methods represent two distinct approaches to enhancing food safety and extending the shelf life of food products. Traditional packaging typically involves the use of inert materials to provide a physical barrier, preventing the entry of external contaminants and protecting food from spoilage. While traditional packaging is effective in reducing physical damage and preserving the initial quality of food, it does not possess inherent antimicrobial properties. On the other hand, antimicrobial nano-packaging incorporates nanoparticles with specific antimicrobial agents, such as silver nanoparticles (AgNPs) or zinc oxide nanoparticles (ZnONPs), into the packaging material. These nanoparticles release antimicrobial agents, which actively inhibit the growth of bacteria, viruses, and fungi, providing an additional layer of protection against microbial contamination. The key advantage of antimicrobial nano-packaging lies in its ability to actively combat pathogenic microorganisms, thereby reducing the risk of foodborne illnesses and spoilage. Unlike traditional packaging, antimicrobial nano-packaging can extend the shelf life of food products by inhibiting the growth of spoilage microorganisms and maintaining food quality over a more extended period (Metak A.M. et al., 2013).

10.7.1 Antimicrobial Properties

1. **Traditional packaging:** Traditional packaging relies on barriers and chemical preservatives to protect food from microbial contamination. These methods provide limited antimicrobial activity and may have drawbacks related to chemical residues and potential migration of substances into the food.
2. **Antimicrobial nano-packaging:** Antimicrobial nano-packaging incorporates nanomaterials with inherent antimicrobial properties, such as silver nanoparticles, zinc oxide nanoparticles, or chitosan nanoparticles. These nanomaterials actively inhibit the growth of microorganisms, providing more effective and sustainable protection against spoilage and pathogenic bacteria.

10.7.2 Extended Shelf Life

1. **Traditional packaging:** Traditional packaging can delay spoilage to some extent, but its effectiveness might be limited, leading to a shorter shelf life for perishable foods.

2. **Antimicrobial nano-packaging:** Antimicrobial nano-packaging extends the shelf life of food products by significantly reducing microbial growth and contamination. The controlled release of antimicrobial agents from nanocarriers ensures continuous protection over an extended period.

10.7.3 Eco-Friendliness

1. **Traditional packaging:** Many traditional packaging materials, such as plastics and metals, are non-biodegradable and can contribute to environmental pollution and waste.
2. **Antimicrobial nano-packaging:** Antimicrobial nano-packaging often utilizes biodegradable and eco-friendly materials, reducing its environmental impact and promoting sustainability.

10.7.4 Real-Time Monitoring

1. **Traditional packaging:** Traditional packaging generally lacks real-time monitoring capabilities, making it challenging to assess the condition of the packaged food during storage and transportation.
2. **Antimicrobial nano-packaging:** Antimicrobial nano-packaging can incorporate nanosensors and indicators that provide real-time information about critical parameters like temperature, humidity, and gas composition. This active monitoring ensures better food safety and quality control.

10.8 EMERGING TECHNOLOGIES AND INNOVATIONS

Antimicrobial food nano-packaging has witnessed significant advancements in recent years, with researchers and industry experts continuously exploring innovative technologies to enhance food safety and quality. Several emerging technologies and innovations are shaping the future of antimicrobial food nano-packaging, addressing current challenges and expanding the possibilities for effective food preservation.

10.8.1 Nanosensors for Real-Time Monitoring

Nanosensors have emerged as a cutting-edge technology for real-time monitoring in various fields, offering unparalleled sensitivity and accuracy. These nanoscale devices are designed to detect and quantify specific analytes or physical parameters, enabling continuous and precise monitoring of critical processes. Some applications of nanosensors for real-time monitoring include:

1. **Environmental monitoring:** Nanosensors play a vital role in monitoring environmental pollutants, such as heavy metals, volatile organic compounds (VOCs), and greenhouse gases. They can be integrated into water and air quality monitoring systems, providing real-time data to assess environmental health and identify potential hazards.
2. **Healthcare and biomedical applications:** Nanosensors have great potential in medical diagnostics and healthcare. They can be used to detect biomarkers, pathogens, or specific molecules associated with diseases, enabling early diagnosis and personalized treatment. Nanosensors can also be employed to monitor physiological parameters, such as glucose levels or blood pressure, in real-time for better disease management.
3. **Food safety:** Nanosensors are increasingly being utilized to monitor food quality and safety. They can detect foodborne pathogens, spoilage indicators, and chemical contaminants, ensuring the safety of food products throughout the supply chain and reducing the risk of foodborne illnesses.

4. **Structural health monitoring:** In engineering applications, nanosensors are employed for real-time structural health monitoring of critical infrastructure, such as bridges, buildings, and aircraft. They can detect stress, strain, and other mechanical properties, providing early warnings of potential failures and facilitating proactive maintenance.
5. **Industrial process monitoring:** Nanosensors are integrated into manufacturing processes to monitor variables like temperature, pressure, and chemical concentrations. By providing real-time data on process parameters, nanosensors enhance process efficiency, product quality, and safety (Khalaf H.H. et al., 2013).

10.8.2 Edible Antimicrobial Coatings

Edible coatings containing antimicrobial nanoparticles are being developed to provide an additional layer of protection on the surface of food products. These coatings can release antimicrobial agents to suppress microbial growth, while also acting as a barrier against external contaminants.

10.8.3 Smart and Intelligent Nano-Packaging

Smart and intelligent nano-packaging represents the next frontier in the evolution of food packaging technology. It involves the integration of advanced nanomaterials and sensing capabilities into packaging materials to create innovative solutions that go beyond traditional packaging functions. Some key features and benefits of smart and intelligent nano-packaging are:

1. **Active preservation:** Smart nano-packaging incorporates nanomaterials with antimicrobial and antioxidant properties, actively inhibiting the growth of microorganisms and delaying food spoilage. This extends the shelf life of perishable products and reduces food waste.
2. **Indicator technology:** Intelligent nano-packaging includes indicators or sensors that change colour or emit signals in response to specific conditions. For instance, time-temperature indicators can show whether a product has been exposed to temperature abuse during storage or transportation, helping consumers and retailers make informed decisions about product safety and quality.
3. **Quality monitoring:** Nanosensors embedded in the packaging can continuously monitor factors like gas composition (e.g. oxygen and carbon dioxide), humidity, and temperature. This real-time data allows supply chain stakeholders to track the condition of the product and make adjustments as needed to maintain optimal quality.
4. **Food Safety:** Smart packaging can detect and signal the presence of foodborne pathogens or spoilage indicators, providing early warning signs of potential contamination and enhancing food safety.
5. **Authentication and Traceability:** Nanotechnology can be employed to create unique identifiers or labels that help verify the authenticity of products and prevent counterfeiting. These identifiers can also aid in tracing the origin and journey of the product through the supply chain.
6. **Interactive Packaging:** Intelligent nano-packaging can interact with consumers through mobile apps or electronic devices. For example, scanning a product's packaging might provide information about the product's origin, nutritional content, or recipes, enhancing consumer engagement and transparency.

However, it is essential to address potential challenges associated with the use of nanomaterials in packaging, such as regulatory concerns, consumer acceptance, and the potential release of nanoparticles into the food. Striking a balance between innovation, safety, and sustainability will be crucial in realizing the full potential of smart and intelligent nano-packaging.

Overall, smart and intelligent nano-packaging offers exciting opportunities to revolutionize the food industry, improve food safety, reduce waste, and enhance the overall consumer experience (Agnihotri S et al., 2019).

10.8.4 Encapsulation for Controlled Release

Encapsulation techniques are being utilized to encapsulate antimicrobial nanoparticles, ensuring their controlled release over time. This approach optimizes the antimicrobial effects and prolongs the efficacy of nanoparticles in combating pathogens.

10.8.5 Combination of Nanomaterials

Researchers are exploring the synergistic effects of combining different types of antimicrobial nanoparticles in a single packaging material. This approach aims to maximize the antimicrobial activity and broaden the spectrum of inhibition against a wider range of microorganisms.

10.8.6 Eco-Friendly Nanomaterials

The development of eco-friendly and biodegradable nanomaterials is a focus of current research, aiming to reduce environmental impact and promote sustainable antimicrobial food nano-packaging. The integration of these emerging technologies and innovations in antimicrobial food nano-packaging has the potential to revolutionize food safety and preservation. However, it is essential to conduct comprehensive safety assessments and regulatory evaluations to ensure the widespread adoption of these technologies and their safe use in the food industry.

10.9 CONCLUSION

Antimicrobial food nano-packaging is an innovative solution to ensure food safety and quality in the constantly evolving food industry. It involves integrating nanoparticles like silver nanoparticles (AgNPs), zinc oxide nanoparticles (ZnONPs), and chitosan nanoparticles into packaging materials. These nanoparticles release controlled amounts of antimicrobial agents, providing sustained protection against microbial contamination and preventing the growth of harmful bacteria, viruses, and fungi. This technology not only improves food safety by reducing the risk of foodborne illnesses but also helps to reduce food waste, which is a significant concern in terms of global food security. In addition, new and innovative nanotechnologies, such as nanosensors for real-time monitoring, edible antimicrobial coatings, and smart packaging systems, are revolutionizing the food packaging landscape. These advancements allow for real-time assessment of food quality, timely responses to potential issues, and provide consumers with reliable information about food freshness and safety. However, it is essential to address challenges related to safety, regulation, cost-effectiveness, and consumer perception to ensure the responsible and successful implementation of antimicrobial food nano-packaging. Comprehensive research, collaboration between stakeholders, and transparent communication are crucial in building consumer trust and gaining regulatory approval for the safe use of nanomaterials in food packaging. Antimicrobial food nano-packaging stands as a promising and sustainable solution to safeguarding public health, reducing food waste, and securing the future of the food industry. As research and innovation continue to advance, this modern focus on foodborne microbes and the integration of antimicrobial properties in food nano-packaging is expected to pave the way for safer, longer-lasting, and higher-quality food products for consumers worldwide.

ACKNOWLEDGEMENTS

I am sincerely thankful to our College and Management, whose expertise and valuable insights significantly enriched the content of this chapter. Their encouragement and enthusiasm inspired us to persevere and deliver our best work.

REFERENCES

Agnihotri, S., & Dhiman, P. (2019). Nanosensors: A brief overview and their applications in environmental monitoring. *Journal of Environmental Sciences*, 75(3):160–180.

Arfat, Y. A., Benjakul, S., & Ahmad, M. (2021). Recent advances in nanomaterials and their applications in food packaging. *Journal of Food Science and Technology*, 58(5):1709–1734.

Bai, H., Jiang, X., & Zhang, Q. (2016). Photocatalytic inactivation mechanisms of TiO2 and Ag-TiO2 against Escherichia coli. *The Royal Society of Chemistry Advances*, 6(38):32080–32092.

Biji, K. B., Ravishankar, C. N., Mohan, C. O., & Srinivasa Gopal, T. K. (2015). Smart packaging systems for food applications: A review. *Journal of Food Science and Technology*, 52(10):6125–6135.

Bintsis, T. (2017). Foodborne pathogens. *AIMS Microbiology*, 3(3):529–563.

Brody, A., Bugusu, B., Han, J. H., Sand, C. K., & Mchugh, T. (2008). Innovative food packaging solutions. *Journal of Food Science*, 73(8):1750–3841.

Carbone, M., Donia, D. T., Sabbatella, G., Serafino, A., Brunetti, V., Amalfitano, A., ... Manno, D. (2019). Zinc oxide nanoparticles as selective antimicrobial agents. *Nanomaterials*, 9(5):743.

Durán-Lara, E. F., Morales-Ramos, A. C., Gómez-Aldapa, C. A., Alcántara-De la Cruz, R., Álvarez-Pérez, O. B., & Rodríguez-Marín, M. L. (2020). Evaluation of antimicrobial packaging and high hydrostatic pressure to maintain postharvest quality of strawberries (Fragaria x ananassa Duch.). *Foods*, 9(10):1368.

Emamifar, A., Kadivar, M., Shahedi, M. and Soleimanian-Zad, S. (2010). Evaluation of nanocomposite packaging containing Ag and ZnO on shelf life of fresh orange juice. *Innovative Food Science and Emerging Technologies*, 11(4):742–748.

Food and Agriculture Organization of the United Nations (FAO). (2011). Food safety and foodborne illnesses. Retrieved from http://www.fao.org/food/food-safety-quality/a-z-index/foodborne-illnesses/en/

Friedman, M., Henika, P. R., & Mandrell, R. E. (2002). Bactericidal activities of plant essential oils and some of their isolated constituents against Campylobacter jejuni, Escherichia coli, Listeria monocytogenes, and Salmonella enterica. *Journal of Food Protection*, 65(10):1545–1560.

Gupta, A., Eral, H. B., Hatton, T. A., & Doyle, P. S. (2016). Nanoemulsions: Formation, properties and applications. Soft Matter, 12(11):2826–2841.

Jafarzadeh, S., Salehabadi, A., & Jafari, S. M. (2020). 10 - Metal nanoparticles as antimicrobial agents in food packaging. In *Handbook of Food Nanotechnology*, ed. S. M. Jafari. Cambridge, MA: Academic Press, vol. 12, pp. 379–414.

Kavakebi, E., Anvar, A., Ahari, H., & Motalebi, A. (2021). Green biosynthesized Satureja recharger Jamzad-Ag/polyvinyl alcohol film: quality improvement of Synchronous s fillet during refrigerated storage. *Food Science and Technology*, 41(1):267–278.

Khalaf, H. H., Sharoba, A. M., El-Tanahi, H. H., & Morsy, M. K. (2013). Stability of antimicrobial activity of pullulan edible films incorporated with nanoparticles and essential oils and their impact on turkey deli meat quality. *Journal of Food and Dairy Sciences*, 4(11):557–573.

Kumar, P., Pandey, G., & Choudhary, R. C. (2019). Applications of Biopolymer in Nanopackaging. In *Biopolymer-Based Nanomaterials in Drug Delivery and Biomedical Applications* (pp. 272–292). Amsterdam, Netherlands: Elsevier.

Li, W. R., Xie, X. B., Shi, Q. S., Duan, S. S., Ouyang, Y. S., & Chen, Y. B. (2011). Antibacterial activity and mechanism of silver nanoparticles on Escherichia coli. *Applied Microbiology and Biotechnology*, 85(4):1115–1122.

Metak, A. M., & Ajaal, T. T. (2013). Investigation on polymer-based nano-silver as food packaging materials. *International Journal of Chemical and Molecular Engineering*, 7(12):1103–1109.

Motlagh, N. V., Mosavian, M. T. H., & Mortazavi, S. A. (2012). Effect of polyethene packaging modified with silver particles on the microbial, sensory and appearance of dried barberry. *Packaging Technology & Science*, 26(1):39–49.

Rai, M., Yadav, A., & Gade, A. (2009). Silver nanoparticles as a new generation of antimicrobials. *Biotechnology Advances*, 27(1):76–83.

Ramos, M., & Valdés, A. (2019). Active and intelligent packaging systems for a modern society. *Trends in Food Science & Technology*, 3(86):173–180.

Ribeiro, M. P., Fuciños, C., Pastrana, L., & Teixeira, J. A. (2019). Nanocarriers for bioactive compounds delivery in food preservation. *Current Opinion in Food Science*, 7(26):18–24.

Sujithra, S., & Manikkandan, T. R. (2019). Application of nanotechnology in the packaging of foods: a review. *International Journal of ChemTech Research*, 12(4):7–14.

Vimbela, G. V., Ngo, S. M., Fraze, C., Yang, L., & Stout, D. A. (2017). Antibacterial properties and toxicity from metallic nanomaterials. *International Journal of Nanomedicine*, 12:3941–3965.

WHO (2015). Word Health Organization. Estimates of the global burden of foodborne diseases. Retrieved 13, 2020, from https://www.who.int/activities/estimating-the-burden-of-foodborne-diseases.

World Health Organization (WHO). (2015). WHO estimates of the global burden of foodborne diseases. Retrieved from https://www.who.int/foodsafety/areas_work/foodborne-diseases/ferg/en/

Index